OPEL
MILITARY VEHICLES
1906-1956

Eckhart Bartels

"Pushing help" for "Opel in field gray": Opel trucks were strategic vehicles for the military.

Translated from the German by Edward Force.

This book originally published under the title,
Opel Militärfahrzeuge 1906-1956,
by Podzun-Pallas Verlag.

Library of Congress Catalog Number: 97-65444.

Printed in China.
ISBN: 0-7643-0267-1

We are interested in hearing from authors with book ideas on related topics.

Published by Schiffer Publishing Ltd.
77 Lower Valley Road
Atglen, PA 19310
Phone: (610) 593-1777
FAX: (610) 593-2002
E-mail: Schifferbk@aol.com.
Please write for a free catalog.
This book may be purchased from the publisher.
Please include $2.95 postage.
Try your bookstore first.

Eckhart Bartels

OPEL Military Vehicles 1906-1956

Schiffer Military History
Atglen, PA

CONTENTS

Front cover: Opel Blitz 1-ton, Type 2.0-12 with 2-liter 6-cylinder engine (36 HP), built in 1938, as an ambulance of an antitank unit, on display at Victory Memorial Museum, Messancy, Belgium. (Photo: Guy Franz Arend, Founder of VMM)

Back cover: Opel Blitz 3-ton all-wheel-drive ls medium off-road truck, Type A, built in 1944, owned by a military collector. In the background the "Uniform Lkw" built under Opel license by Mercedes-Benz, Type 701, built in 1945. The monogram on the grille was not factory-mounted. (Photo: Siegfried Maier, Aalen, Württemberg)

Foreword

This book documents the history of the Opel firm in connection with the military as an orderer and buyer as well as in terms of influence and political power.

The Opel works, in their 125 years of history, were always a part of their historical milieu. As a family business, as also later as a stock company, Opel always had the goal of being able to offer good quality at moderate prices through large-scale production. Thus Opel ranked among the best in the market segment of its time. These qualities were particularly valued by the military in wartime too.

Here for the first time, the reader finds a thorough presentation of the Opel firm's history in connection with the military. The basic information comes from the archives of the Adam Opel AG, well-known sources and my own archives, which could be completed with new details thanks to the help of several old friends from Opel. Essentially, I was concerned with technical developments and model histories, which are often of interest to scholars of automotive and military history, collectors and model builders.

Many friends from the Alt-Opel-Interessengemeinschaft cooperated in assembling this book, and I thank them most heartily for their assistance. To represent them, I would like to mention only the former member of the Opel publicity department Bodo Fischer, as well as Hans-Joachim Raab and Carsten Messer. They showed me archival material that was previously unknown or had not been published in this context.

Eckhart Bartels
Ronnenberg, Autumn 1994

Opel bicycles in paramilitary service, 1935.

Adam Opel's Factory: Sewing Machines for Uniforms and Bicycles for the Troops

It was 1862 when the firm that is now the Adam Opel AG was founded. It was the same year in which Otto von Bismarck was named Prussian Minister President and Foreign Minister by King Wilhelm I. He provoked three military conflicts, culminating in the Franco-Prussian War of 1970-71. A time that cried out for soldiers in uniform moved more and more tailors and factories to call for sewing machines. At that time, Adam Opel had finished his journeyman years and assembled his first sewing machine in a Rüsselsheim cow barn. These were sewing machines that still required sturdy cast iron parts from France. The war of 1870-71, of course, influenced, though temporarily, the production of sewing machines, which had been off to a good start, but soon improved models were able to liven up the market.

For Adam Opel, the war with France was quickly forgotten. Since his years of apprenticeship in Paris, the firm's founder had always maintained a special fondness for France, and shortly after the war ended, he was able to resume his intensive trading activities. On one of his business trips to Paris, he got acquainted with the first British bicycles. From 1886 on, these bicycles were developed further and produced as real "racers" at his factory in Rüsselsheim on the Main. More than 16,000 left the small city on the Main around the turn of the century. The bicycle became a popular means of transportation, and Opel became the biggest bicycle producer on the European continent.

The first Opel sewing machine of 1862.

Bicycle advertisement, 1902: Transporting goods to Kiaochau.

The German military authorities had also recognized the bicycle as an important tool and assigned particularly robust designs to the military bicycle riders in individual troop units. First envisioned only for use by orderlies, they were also used for patrols and by intelligence units in the large-scale maneuvers before World War I. In the last years of peace, this helped Opel's production to expend to over 38,000 units. During the war years, production of touring bicycles stagnated at 10,000 a year.

In addition to the traditional service arms and types of troops, long columns of bicyclists traversed the roads during the days of World War I: infantry on bicycles, Jäger bicycle units and bicycle companies. Often the men were trained riders and sportsmen who controlled "their machines" well, whether in the front lines of battle or building or replacing "telegraphic connections." Thus the bicycle, and particularly its "pneumatic" tires, established themselves as articles important to the war. At home, bicycling was banned, as bicycles and tires were requisitioned for the troops. Exceptions to this ban were granted, particularly for the many invalids who had been sent home, and who could use special bicycles or tricycles.

1916: A naval bicycle company on the advance in Flanders.

At this time, the first experiments were made with folding bicycles, which the international manufacturers had already displayed at the world's fairs in Paris and London, but which had attracted practically no notice in Germany. Now the French Army also mobilized its soldiers on small, handy, easily carried bicycles. But these special folding bicycles were not decisive in battle. Because of a lack of workers and raw materials, Opel soon had to give up testing folding bicycles. To the military authorities, it was more important to meet the urgent need for army trucks, and Opel had meanwhile become one of the most important military suppliers, which delivered over a thousand trucks of the three-ton class per year.

The First Trucks: Testing in Reliability Trials

Opel's entry into automobile building took place in 1899 with the acquisition of the Lustmann motor vehicle factory in Dessau. The designs of the former court locksmith, though, proved to be of little use, so that cooperation with Renault and Darracq was soon sought; an accord with Darracq ran successfully until 1909. In 1902 the sewing-machine and bicycle factory took part in the Hamburg Auto Show with its first original design, the Opel 10/12 HP motor wagon. Just two years later, Opel also offered its first commercial vehicle, an "Advertisement Vehicle" based on the 10/12 HP two-cylinder model. The "Opel-Motor-Lastwagen" had a carrying capacity of 12 to 15 hundredweight of payload. It was available from the body shop in Rüsselsheim with either open or box body. While these trucks cost between 6000 and 6500 gold Marks, the first buses for eight passengers, depending on their equipment, cost up to 2000 Marks more. Higher prices also had to be paid for more powerful 30 HP four-cylinder motors.

The sale of commercial vehicles began only slowly. In the first year eight trucks left the town on the Main, in the next three years 17, 26 and 24 were sold. The teamsters' level of society preferred the kind of horsepower produced by oats. The military agreed with them. Mobility for the troops was guaranteed only by the storekeeper sergeant with his draft horses, and the automobile was regarded as not dependable enough. Thus in 1907 the licensing statistics kept in the German Reich showed only 957 commercial vehicles, of which only 75 bore the Opel monogram on their radiator grilles.

Since experience with solid commercial vehicles was lacking, all the manufacturers took part gladly in the numerous reliability trials and winter tours, to be able to recommend their products. The military too, courted as a contract giver by all the builders, was also there, represented by observers. But national goals were not the only ones. The international competitions placed particularly high demands on men and machines. In particular, they included the Russian tours, which could only be participated in by good teams and reliable automobiles. The Russian military was in search of suitably transport vehicles for its troops and wanted to use the lure of sport to gain an overview of what trucks were on the market. The competition for commercial vehicles was under the patronage of the Tsar. The "General Automobile Journal" of August 11, 1911 reported on the Russian Tsar's competitive tour and the "glittering debut of the new OPEL three-ton truck", which had covered the Petersburg-Moscow-Petersburg run of some 2000 kilometers in "faultless style and without any defect." Only two-thirds of the route was run on Russian roads, the rest on field paths in considerably worse condition. The victorious vehicles also won the admiration of the Russian War Ministry. Immediately after the Tsar's Prize was awarded to Dr. Ludwig Opel at Yalta, a number of Opel three-ton trucks were ordered for use by the Army. Thus the Russian Army also became a source of experience with German trucks.

The Tsar receives Dr. Ludwig Opel in Crimea at the end of the Tsar's Prize Tour "Petersburg-Sevastopol", 1911.

State Support: Subsidies for Trucks

What the reliability trials contributed to the technical development of the automobile was amplified by the state subsidies, which increased the demand for truck types that could be used in war. The automobile, though, still played a subsidiary role in the plans of the Army command. As a means of transporting troops and military equipment over long distances, the railroad dominated. Therefore, the state did not want to invest much capital in trucks for military use, especially as the technical developments changed with every automobile competition. Thus in 1908 the Army command got the idea of financially subsidizing trucks and tractors for use in the field, meaning towing tractors, towing trucks and trailers. Every buyer of a truck recommended by the Army for subsidizing received a one-time contribution of 4000 Marks and a further yearly payment of 1000 Marks for the next five years. Thus more than half the purchase price of a typical truck was paid. The buyer, thus obligated, had to keep his truck ready at all times in a technically faultless condition, which was checked annually by the military officials.

Along with these regular checks and condition reports, the Army command also promoted "subsidy tours" in order to be able to have the trucks' capability, sturdiness and economy judged by military observers. These truck test drives covered several thousand kilometers and exposed design faults mercilessly. All the same, almost all the manufacturers of trucks applied to participate, since they wanted to secure for themselves a part of the buyers' demand for subsidized trucks. In 1908, when the new law took effect, only four manufacturers had fulfilled the Army's requirements for subsidized trucks: Benz, Büssing, Daimler and NAG. The trucks weighed up to nine tons in all. Through better designs, the requirements for the "standard type" of the German Army were rewritten in the summer of 1911 and limited to a maximum weight of four tons. Again it was Büssing that was first to have a truck that met these requirements in production. On November 10, 1911, such an army truck and trailer set out on a 2000-kilometer test run, which was then evaluated by the Technical College in Charlottenburg. The subsequent Army orders from all the important truck factories were based on these results.

Other than these heavy trucks and trailers, a smaller number of three-ton trucks also proved themselves. To its earlier offerings of light trucks, Opel had added a truck it had developed with a three- to four-ton payload. The Type 16/35 HP with two-cylinder engine (100 mm bore and 130 mm stroke) had a top speed of 25 kph on hard-rubber tires, and even reached 20 kph with a full load. The chassis alone, with front fenders, was available f.o.b. the factory for 12,500 Marks. The various bodies could be had from Rüsselsheim with the stylish Torpedo streamlining on the cab. For such a canvas-back body, one had to pay another 1800 Marks, while the also available omnibus body cost an extra 5500 Marks.

Since the Army command also promoted standardized trailers financially, trucking companies could order complete truck-and-trailer sets from the Opel works at Rüsselsheim. Such a uniform trailer with

Opel truck chassis for 3- to 4-ton payloads.

Kaiser Wilhelm also used an Opel Landaule Form S of Type 16/35 PS, in 1911.

hard-rubber tires, and with a cover, cost 3900 Marks; the simplest version, with iron tires, only 2500 Marks. With the introduction of this standard trailer, commercial vehicle production doubled at the Opel automobile factory, reaching 150 units a year.

At the beginning of 1913, there were already 825 subsidized trucks in use in various businesses. The breweries dominated, with a share of 43% of all the licensed trucks. In second place, with 23%, was the transport business, followed by the brickmakers, mills and farmers. This was no different from their military uses, with only a few jobs like carrying mail and serving as ambulances assigned to motor vehicles at first. Only during the course of World War I were other types of use added. The cavalry divisions used more and more of the fast and handy three-ton trucks, which were also known to the troops as cavalry trucks. Later, the truck was the only means of transportation used by the newly formed flying units on the ground.

An Opel road train of Opel truck and trailer, 1911.

Opel aircraft engine of 1912.

Opel trucks in action with a flight formation.

The Automobile in the War: Driving Schools for Motorized Battalions and Black Corps

Around the turn of the century, the military also became aware of the automobile. In 1899 the first motor droschke appeared in Berlin, and in the autumn of the same year, the first automobile exhibition opened its doors on the Karlstrasse in Berlin. At that time there were only incomplete trucks, whereas the passenger car had already attained a certain maturity. It was the time at which the Imperial Automobile Club (K.A.C.) was founded, which had a decisive effect on the beginning of motor vehicle use in the German Army.

After the Kaiser's Maneuvers of 1899, in which several K.A.C. members had taken part, it was agreed in a contract that in the case of mobilization, vehicles and drivers would voluntarily place themselves at the disposal of the military command. On March 11, 1901 the Army command decided to establish a "Self-driver Command" under the leadership of a captain in the testing department of the transportation troops in Berlin. At first the teams were taken over by the railroad troops and trained as driving personnel. This early phase was evaluated as particularly successful, since the first requirements for military trucks were originated at that time, which became generally applicable in Prussia in 1908, Bavaria in 1909, and then in Austria-Hungary as well.

On April 1, 1907 the formation of the test unit was completed and the 150-man-strong "1st Motor Vehicle Unit of the Transport Troops" at Tempelhof Airfield had the assignment of training new military drivers. Thus the first German special unit for military motor vehicle use expanded, and four years later, on April 1, 1911, it was subordinated to the newly formed "Inspection of the Military Air and Motor Transport", ILuK for short. After September 11, 1911, empowered by the War Ministry, it was allowed to award the coveted "Fahrerlaubnisscheine" driving licenses. The tested student drivers received not only a uniform but also the black leather suit required for military driving and a golden "K" on the shoulder

Russian Tsar's Prize Tour, September 20-29, 1911, with three Opel cars of Dr. Opel, Michel and Jörns.

pieces. The thus identifiable "Black Corps" saw service at the beginning of the war as field mail drivers or in motorized ambulance units ("Sanka"). But they also filled the troops' needs for supply transport, with meat wagons or field kitchens, as well as in the new airplane or dirigible units. The "Black Corps" had become needed helpers almost everywhere.

In August 1914 the field motorized forces consisted of 114 mobile vehicle formations with about 200 officers, 8000 soldiers and 4000 vehicles, including those assigned to the command staffs and special weapons units. During the course of the war, more and more motor vehicles were found in all service arms, including the artillery and "mine-launcher weapon" units, as well as the mobile communications troops, which were equipped with radio stations in special vehicles.

Thus the unit's motor pool met its needs for manpower, vehicles and fuel. To be sure, the consumption of materials at the fronts was greater than the amount of needed equipment that could be produced in Germany. The capacities of about a dozen and a half factories that made subsidized trucks at the beginning of the war were fully utilized, and had meanwhile been augmented by another ten manufacturers of military vehicles.

The Opel works, thanks to large contracts from the procurement offices, had concentrated production on the three-ton trucks, of which almost 100 units per month could be delivered. But about forty new smaller cars for messenger service and scouting troops also reached the military every month. In everyday military life, the Opel "Puppchen" of Type 5/14 HP, already very popular in peacetime, proved itself very well. It was found, used as a courier car, in all sectors of the front.

But Opel had its engine-building operations going at full blast, along with its motor vehicle production. The four-cylinder aircraft engine with cylinders cast in pairs, which originated during the war, pro-

From a private album: An Opel with the staff of the 208th Infantry Regiment.

duced 60 to 70 HP at 1250 rpm, but proved to be too weak to stand up to everyday wartime uses and had to be replaced during the war by an Argus six-cylinder aircraft engine built under license, and later also by a BMW. In 1915, as a replacement for their own weak four-cylinder type, they developed a six-cylinder aircraft engine that could run at speeds up to 1450 rpm. One year later, this powerplant was mounted experimentally in a car, which thus became the first six-cylinder Opel car. This Opel car, designated the Type 18/50, was built only in small numbers as an officer's car. The ancestor of all Opel large-vehicle engines produced up to 56 HP and displaced 4705 cc. Over 100 such machines were built every month as aircraft engines and delivered to the aircraft-building industry.

The trained drivers and auto mechanics also found a further sphere of influence in the staff motor pools. Here they had the job of maintaining the mobility and readiness of the motorized units. Their work at these motor pools was not easy. A day began at 6:00 A.M. After coffee, they had to drill for half an hour. Only then did their chief duties begin: They repaired motor vehicles or towed them in from the supply lines. A large bus held their writing room. The repaired vehicles and spare parts were accounted for bureaucratically. Ten to fifteen soldiers, with a column leader, devoted themselves to repair work, broken only by breaks and a group lunch. The daily work at a motor pool lasted until 6:00 P.M., or sometimes only at 7:00 when they fell in to receive orders.

The staff motor pool was set up behind the front. Preferred locations were factory premises, in the shelter of which the mobile facilities could be at least partly concealed beneath large tent canvases. There was a smithy for axle repairs, a carpentry shop for body work, a glass works, a plumber's shop, saddlery, tailor shop and even a shoemaker's shop. One particular work column overhauled engines, as the overworked motors often needed new bearings. At a rub-

In the test run for light trucks, run by the War Ministry on March 27, 1912, a 1910 Opel 5/10 HP car is loaded on a flatbed truck.

Garage of the "Automobilrotte" of the Russian War Minister, March 1912.

The motor pool for the staff of Field Marshal von Hindenburg in the East.

ber depot, covers and inner tubes were patched. The necessary artificial light for the facility was provided by a generator truck, whose dynamo also provided power for the electric motors in the repair shop. A motor pool supplied the army with fuel, oil, tires, spare parts and tools. To always have enough materials on hand, the transport columns had to be underway constantly. The defective or shot-up vehicles were collected so that everything usable could be removed from them. The rest remained at a "graveyard" near the motor pool until rail transport to the factories back home could be arranged. Thus the staff motor pool had an extensive storehouse of raw materials, spring steel, iron and brass on hand for its own use as well as for the mobile repair shops. These mobile facilities for smaller repairs followed the motorized troop units directly, to be able to make repairs on the spot. But the longer the war lasted, the shorter became the supplied of raw materials and parts. The supplies of gasoline at the beginning of the war amounted to 76,000 tons, but they were used up quickly. Since supplies from Galicia and Romania could not be had, the production of benzol in Germany was begun anew. For that reason, the Quartermaster General set up guidelines for the motorized units as early as 1915. They included the limiting of traffic in Germany to purely service-related purposes and the requisitioning of additional motor vehicles for war service. This in turn increased the use of fuels, which could not be replaced, even by benzol. Thus the supplies of fuels that could be mixed with benzol also came under the military command. A shortage of rubber for tires also developed when the war began. Because of this, the combat troops soon received, with few exceptions, trucks with metal tires, which made their use considerably harder. At the same time, anti-aircraft guns mounted on trucks (Kflaks), with four-wheel drive for greater mobility and readiness to fire in any direction, were developed as a barely suitable defense against the enemy's tanks. Small-caliber guns for cars and field guns mounted on trucks were also used in the last years of the war.

Field postcard of 1916: In front is an Opel 10/24 HP of 1912, in the background a vehicle with a bar to ward off rope barrages.

The first armored Opel of 1905, which was planned as a "battle-wagon" for higher officers. For lack of interest, it never went beyond this one example.

A big Opel Type 25/55 HP, on duty in Gallipoli, Turkey, on September 16, 1915.

Opel did not take part in these developments. To be sure, in 1905 an armored vehicle for high-ranking officers, on an Opel-Darracq chassis, has been introduced, but it did not interest the Army command and thus was not developed further by the automobile factory. That was the situation with vehicle suitability for military use. Despite considerable limitations because of manpower and material shortages, over 4600 army trucks, 2400 cars, 52,000 bicycles and more than 4000 aircraft engines were produced by the Armistice of November 11, 1918. When the French occupied the Rüsselsheim factory in December 1918, they confiscated completely worn-out tools and machines. The end of production halted the forced labor of 600 prisoners of war. 4515 Opel workers were out of work. 191 of their colleagues had died in the war. They were among the almost eight million people to whom World War I had brought only death.

A truck column to transport meat, at the market place in Anizy, France.

Toward the end of the war in Rumania: Drivers rebuilding a kitchen truck on Opel chassis.

A kitchen truck of a motorized field unit.

Drivers rest by their Opel trucks.

An Opel 10/24 HP as "darkroom truck" for photography, also usable as an ambulance.

An Opel bus in military service, Serbia, 1916.

The Opel Regelwagen: No more orders from the Reichswehr

The motor vehicles used in World War I had much in common with prewar civilian vehicles. Changes for wartime use were made in building engines and in strengthening the hard-pressed drive trains and running gear. Armor plate, special bodies and special military developments for the motor vehicle were scarcely found at all. An armored vehicle, contracted for by the Army command after its bitter experiences on the western front, could not be put into production by the German steel industry as of mid-1918.

When the war ended, the military ceased to award contracts to Opel. The factory itself was occupied by French soldiers. Material shortages, worn-out or dismantled production facilities and increasing inflation considerably limited automobile production. Only one-fifth of the 40,000 motor vehicles delivered by the end of the war had survived long-term action. Over 4000 of the trucks used in the war bore the Opel badge on their grilles, but only a few were still usable after the return of the German troops. Trucks that could carry cargoes to meet transport needs in postwar Germany were lacking. Thus Opel continued the production of three-ton trucks, though in very reduced numbers. At first this was done against the will of the occupying powers, for "export" from French-controlled Rüsselsheim into occupied Germany contravened the customs and trade regulations applying to the 50-kilometer-wide demilitarized zone. Despite that, 582 cars and 572 vans and trucks were assembled and sold in 1920.

The three- or four-ton Opel trucks that were built in Rüsselsheim beginning in 1920 were descendants of the standard trucks built during World War I. The standard truck, under the guidelines of the Normalization Commission, had to meet strict standards that had been derived from experience with the various types used in previous years. The requirements were developed by the vehicle technical test commission of the military high command:

– Almost all parts coming from various factories and exposed to wear had to be interchangeable.
– All screws, bolts and nuts had to have the same threading.
– Front and rear wheels, plus gears and chains, had to be interchangeable among brands. Likewise the radiators.
– The shift pattern had to be the same in all models.
– Dimensions for bodies had to be uniform.

Opel delivered the open rear body with dropping sides from its own factories. A canvas cover was held by four removable hoops. The cab was enclosed by leather at the sides, with only the rigid roof and removable front panel made of metal. Fuel consumption for the Opel Regelwagen with a three- to four-ton load for a distance of 100 kilometers was stated as between 32 and 35 liters of benzol. The power came from a 4-cylinder, 40-HP engine with a bore of 115 and a stroke of 156 mm.

Along with the Opel 0.5-, 0.75- and 1-ton delivery vans that were also available (1919-1920 series) and the Opel trucks made for 1.5-and 2-ton payloads, the 3-ton Regelwagen was built in the uneasy years after World War I, until 1924, with about 2000 built in all. After the end of inflation and the introduction of the Rentenmark currency in December 1923, the sales of trucks remained unsatisfactory. Along with the first Opel "Laubfrosch" produced on a new assembly line, only 154 trucks and delivery vans were sold. Only then came a new generation of small delivery vans and medium trucks for civilian use. Production of the chunky Regelwagen was halted, for even the military no longer called on Rüsselsheim for them. In any case, in 1926 the Reichswehr had wanted better off-road capability for military vehicles, and had turned the development of three-axle trucks over to other manufacturers. Thus the Opel works could be devoted completely to the new market for light vehicles. This was done with success.

This situation showed that the most notable identifying mark used to tell the individual types apart, the radiator, remained the same on all Regelwagen trucks. The tubular radiator had the advantage that, after unscrewing the upper water chamber, tubes that had come loose could be replaced easily. Thus repair shops and vehicle owners could carry out the cleaning of the cooling systems much more simply.

A restored three-ton Opel truck of 1915 (Sinsheim Museum).

An Opel project picture of the Regelwagen, 1920.

The propulsion via driveshaft and differential, the "shaft drive", was also available, along with the old established chain drive. The advantages of the low-wear shaft drive could not be utilized for all vehicles on account of the use of iron wheels and the resulting hard shocks. Normally, then, the chain drive well known from prewar days was marketed, though it was prone to more dirt and wear. Here the power was transmitted by change gears, not directly to the rear axle, but through a two-link shaft mounted ahead of it. Two roller chains then carried the power to the large toothed discs attached to the brake drums of the rear wheels.

The design principle of the Blitz fast truck was successful from the start: Low gross weight along with high payload capacity led to its use in almost all branches of commerce. At the international auto show on the Kaiserdamm in Berlin, the Opel displays were constantly surrounded by crowds. Along with the new 1.8-liter, six-cylinder cars, which awakened the particular interest of Reich President von Hindenburg, Opel displayed the new 2.6-liter 4-cylinder 1.5-ton and the two-ton Blitz trucks with 3.5-liter 6-cylinder engines. The tour and bus business people also admired an unusually nice-looking 3.5-liter Blitz bus that carried 24 people.

OPEL Truck

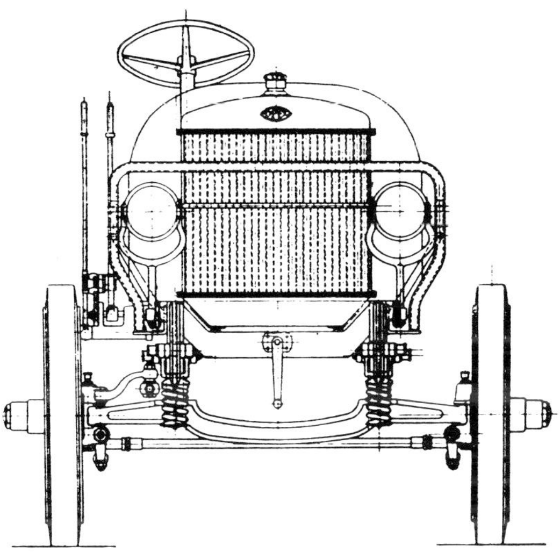

Despite the bad economic situation, the competition had retrogressed further, and the new Blitz could compete well against its competitors with a 17% price increase since before the war. In the summer of 1931, a two-ton truck with a long chassis and a 2.5 ton type were added to the program. With them, the Adam Opel AG, through its extensive network of dealers, expanded its truck sales in all areas. This was based on their customers' buyer analyses for handworkers, truckers, breweries, department stores, bus lines and communities. The latter utilized ambulances, fire trucks and service vehicles. Only one target group was missing: the military. At that time when over six million were unemployed, the Reichswehr concentrated its meager financial means on the acquisition program begun in 1928. Opel trucks were not included in it. In the end, Opel had not applied for development or manufacturing contracts, yet the management in Rüsselsheim was interested in motorizing the broadest levels of the population.

Only after the National Socialists came to power and established their employment program, was the economy's demand brought back to life, which caused Opel's monthly production to increase to 11,500 cars and trucks by March 1935. At that time, capacity limits were reached for the first time, which led the firm to consider a second manufacturing plant.

OPEL

Opel-Dreitonnen-Lastwagen

Regelwagen nach den Richtlinien der Normalisierungskommission

Front page of the 1920 sales brochure.

The New Truck Business: From the first "Blitz" to the Brandenburg Factory

Now that the military no longer gave the firm contracts for trucks, Opel's offerings were chiefly delivery vans and small trucks derived from its car program. This was to change only when, in 1929, the American automobile firm of General Motors bought first 80% and then the rest of the firm, which had become a stock company, Adam Opel KG. Experienced in worldwide business, the marketing strategists of GM recognized that in Germany the truck market could be expanded. For that reason, early in 1930 the firm's car and truck production became separated through the establishment of a new division. Opel adopted the concept of the Chevrolet truck and the production tools of the Buick-Marquette motor. The two-ton truck grew into a new truck type, with which Opel was to be very successful in the thirties.

For the 1.5- and two-ton trucks, the Opel advertising department had to find an internationally understandable product name as a means of encouraging sales. For this, the device of a prize contest was chosen. Awaiting the winner were a new 4 HP sedan as well as four 500 cc Opel Motoclub motorcycles, from the production of the former factory in Brand-Erbisdorf, Saxony. More than 1.5 million entrants took part in the name contest. On November 24-25, 1930 the sales manager P. Andersen, during the course of a dealer convention in the big Ufa Palace in Frankfurt am Main, introduced the new name for the 1931-issue Opel truck: OPEL-BLITZ. This name had already been used by Opel for a fast man's bicycle built around the turn of the century.

In addition, vehicles equipped with a special rear axle transmission could also be used as towing tractors for 5-ton semi-trailers.

Successful entry into the market: A week's production of the new 1931 Opel Blitz trucks ready to be delivered from Rüsselsheim by drivers.

Spezial Kastenwagen für Großstadtbetriebe wie Zeitungen, Warenhäuser usw.

Spezial Kastenwagen für Wäschereien, Reinigungsanstalten usw.

Spezial Vieh- und Fleischtransport-Kombinationswagen

SPEZIALAUFBAUTEN für jeden Betrieb und für alle Sonderzwecke

Die starken, widerstandsfähigen Blitz Fahrgestelle eignen sich hervorragend für den Aufbau von Spezialkarosserien, wie sie von wichtigen Zweigen der Industrie und des Handels zur wirtschaftlichen Gestaltung des Transportdienstes benötigt werden oder wie sie sich für öffentliche Betriebe als zweckmäßig erweisen.
Auch in diesen Sonderfällen sprechen vor allem Gründe der Rentabilität für die Verwendung der Blitz Fahrgestelle. Die Preise dieser Fahrgestelle — mit und ohne Führerhaus — sind so niedrig, daß in vielen Fällen die Kosten für ein Blitz Spezialmodell geringer sind als die für Normalausführungen anderer Fabrikate. Der günstige Radstand ermöglicht den Aufbau von Karosserien mit praktischer Raumausnutzung.

BLITZ

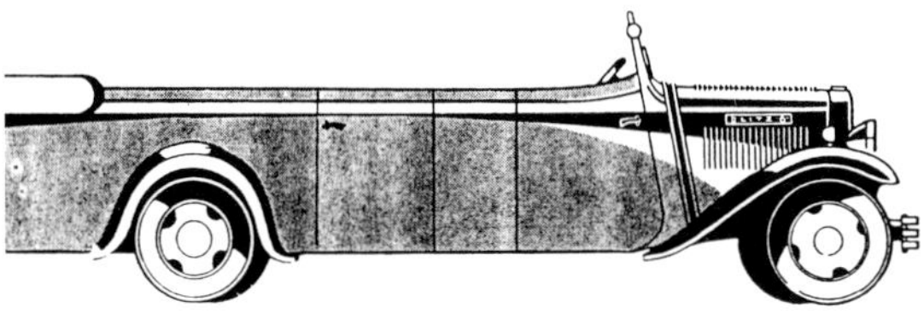

Spezial Omnibus für 24 Fahrgäste

5 to Spezial-Sattelschlepper

Spezial Möbelwagen mit besonders langem Radstand

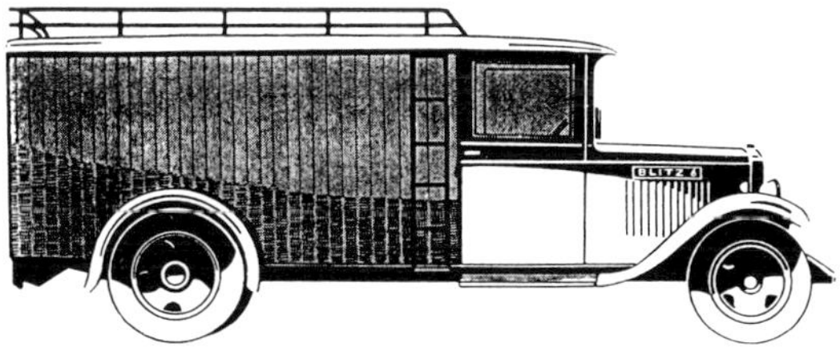

Spezial Stabholz Kastenwagen für Warenhäuser und ähnliche Betriebe

Spezial Fleischerei-Muldenwagen

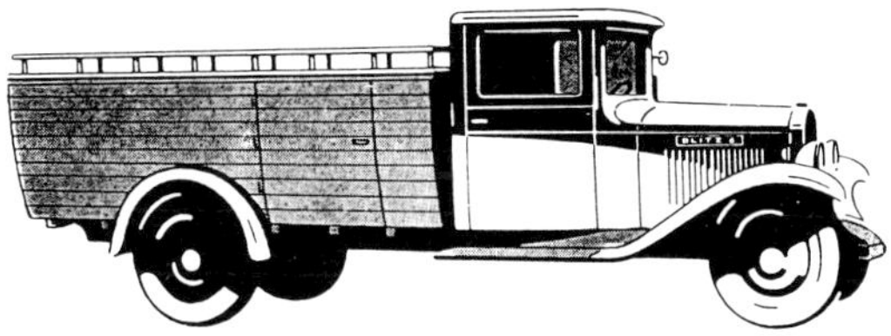

Spezial Flaschentransportwagen für Brauereien, Mineralwasserfabriken usw.

Spezial Omnibus für 12 – 16 Fahrgäste

Spezial Omnibus für 28 – 30 Fahrgäste

2 1/2 to Pritschenwagen mit Vierradanhänger

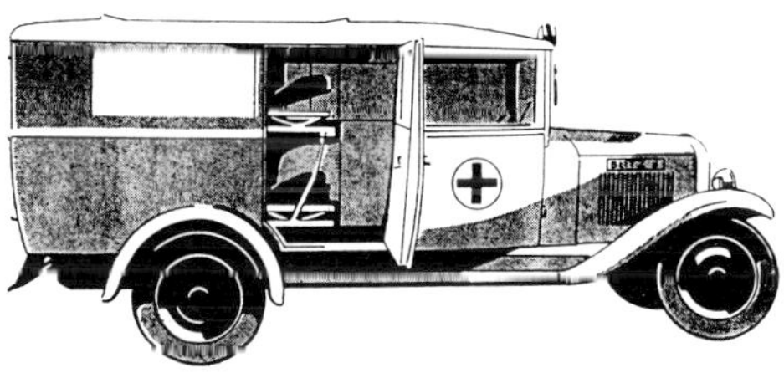

Spezial Krankenwagen

An Opel Blitz for every customer, but not for the military.

The Opel Truck Program in 1935:

Blitz 1-ton 2-liter 6-cylinder as chassis, chassis with cab, box or open body

Blitz 2-ton 3.5-liter as chassis and open truck with 3.41- and 4-meter wheelbase

Blitz 2.5-ton 3.5-liter as chassis and open truck with 4-meter wheelbase

Blitz 2.5-ton 3.5-liter special chassis, such as for buses, with 4.65-meter wheelbase.

On September 7, 1935, five Opel Blitz caravans began a tour of Germany as a mobile sales show. There were eleven Opel Blitzes, all painted red, each with a different body, led by a loudspeaker truck.

Opel Blitz 2.0- or 2.5-ton truck, 1931-1933.

New small Blitz trucks at the Opel sales display in Celle, 1935.

Autumn 1935: The new truck factory in Brandenburg rises.

Spring 1936: View of the factory from across the Silo Canal.

This was a good starting basis for a truck type that in 1936 became the first German truck to correspond to the requirements of the Reich Transport Ministry for production trucks with off-road capability.

On April 1, 1935, the management of the Adam Opel AG decided to build a new truck factory in Brandenburg on the Havel, and twelve days later the first shovelful of earth was turned in the presence of high military brass. The development of an improved Opel 3-ton truck was almost finished and required new production facilities for economic large-series manufacture. Opel purchased a plot of land in the northwest part of the city of Brandenburg, measuring 340 Prussian Morgens, or 850,000 square meters. The new Opel factory was to be built in only 190 work days between Hohenzollern Street and the Silo Canal. 1200 men worked day and night, and after 70 work days the steel frame of the main hall was already standing. On October 16 the complete raw structure was done. Just a month later, the first fifteen Opel Blitz trucks to be built in Brandenburg rolled off the assembly line. On January 7, 1936 the new factory was officially opened by the Reich Transportation Minister, Baron Eltz von Rübenach.

Not only did the planning for the construction schedule work out successfully, but the design of the assembly halls was also a masterpiece of rational planning. In a gigantic, light-bathed assembly hall of 178 by 136 meters, all the work processes from raw material to the finished truck were unified. Freight barges brought fuel by canal, and a branch line of the German Reichsbahn enabled materials to be unloaded and stored right at the factory. On thirteen parallel rows of machines, the crankshafts and camshafts, cylinder blocks, gearboxes and frames, front and rear axles were made. The pieces moved on 27 fully automatic transport belts with a total length of five kilometers to the assembly line, from which, since the beginning of production there, fifty trucks rolled in every eight-hour working shift.

The factory began with a crew of 680 employees, but this number more than doubled by the end of 1936, allowing a yearly production of about 20,000 Opel Blitz trucks.

The factory opens, January 7, 1936: Reich Transport Minister Baron von Elta-Rübenach, Plant Manager Dr. R. A. Fleischer, and Corps Führer Hühnlein with the first 1.5-ton Blitz.

One of thirteen rows of machines reworking the raw material into parts ready to install, such as the rear axles shown here.

One day's production awaits test driving and shipping out.

The assembly line at the end of 1936, with the first new 3-ton truck (second vehicle built).

The 1938 Auto Show: From the P4 Geschäftswagen to the Blitz with 1.0-, 1.5- and 3-ton payloads.

The "S" Trucks: Military Requirements for Off-Road Capability

"Few types – large numbers" was the motto of the Adam Opel AG. This principle was also put to good use in truck construction. Thus new designs could become reality and improvements could be made in series production more quickly than at other firms. At the very beginning of the thirties, Opel, with its Blitz fast truck, had already come onto the market with a progressive new "light truck." Despite its 2.5-ton payload, the chassis had a very low inherent weight of 1550 kilograms. These design principles, which also corresponded to the military concepts for medium trucks, were also accepted by the civilian customers after some initial mistrust. The Opel 2.5-ton truck offered the best qualities for a truck with limited off-road capability. After valuable encouragement from the Reich Transportation Ministry, Opel developed the Blitz design concept in further experimenting and constant testing, so that the technical values, almost without modifications, also served as guidelines for the entire German automobile industry.

On February 28, 1935, in the second law to change the motor vehicle tax laws, the requirements for off-road-capable production trucks were made known, incorporating considerable improvements for owners of commercial vehicles. One year later, the Reich Transportation Minister announced the design requirements for regularly marketed trucks with limited off-road capability. They included:

1. Good relationship of inherent weight to payload.
2. Good payload weight (based on allowable total weight).
3. Trouble-free engine functioning when the vehicle was tilted lengthwise or sideways.
4. At least five forward speeds with off-road gears.
5. Good ground clearance.
6. Little overhang (large off-road angle).
7. Much axle-limiting capability.
8. Large-size rims and tires.

The prototype of the new three-ton generation, with license plates from the Hessian province of Starkenburg.

Introduced at the 1937 IAMA in Berlin: the Opel Blitz three-ton "off-road capable", still with its forerunner's 3.5-liter engine.

The new 3.6-liter engine followed in the late summer of 1937.

The first truck recognized as "off-road capable" was the new 3-ton Opel Blitz "S" introduced at the end of 1936, at first still with the 3.5-liter, 64-HP engine, which was replaced in 1937 by the new 3.6-liter, 75-HP engine. With that, the Opel Blitz "S" became the standard truck of the third Reich. The type designation "S" stood for "series" or "steuerermässigt" (lowered tax) in reference to the subsidized trucks of World War I days. This Opel truck thus received, for the first time, a 33.3% tax reduction. Its yearly tax was only 252 Reichsmark instead of the regular 378 RM.

Along with fulfilling the tax requirements, other features resulted in its immediate sales success: oversize 190 x 20 tires, strong axle limitation, five-speed gearbox with off-road gear for increased climbing ability and better traction on hilly field and wood roads, as well as a hitherto unattained ratio of load to empty weight. The new Opel Blitz, with a total weight of 5700 kg, could carry a payload of 3260 kg, making 133% of its empty weight of 2440 kg. The rest of the German commercial vehicle industry followed this pattern a year later: Büssing-NAG with its Type 654 as well as the Model 30 Burglöwe in the three-ton class. Hansa-Lloyd, Ford, Henschel and Magirus also offered 3-ton S trucks as of 1937. The MAN Z2 with its 3.5-ton payload (1938) and the Mercedes-Benz LGF 3000 with its three-ton load (1939) came on the market only later. Meanwhile, Opel trucks had already gained 35.6% of the market in the three-ton class, an advantage that no competitor could overcome.

Opel 3-ton S trucks proved themselves in all areas during prewar times, whether at construction sites on the Reich Autobahn, in gravel pits, or in agriculture and forestry. Their first militarily organized activities during the building of the West Wall confirmed the decision to acquire the three-ton Blitz S for Wehrmacht use. In 1937 there were already 1256 of these trucks delivered to the Wehrmacht. There were also numerous exports to South American governments that equipped their troops or paramilitary units with Chevrolet or Opel Blitz trucks.

The production version of the three-ton Blitz, Type 3.6-36.

A hard test for an Opel 1.8-liter on the DDAC 1935 Winter Tour at the Titisee.

"Three-Day Harz Tour", June 14-16, 1922, with four Opel 1.8-liter cars and a two-ton Opel Blitz.

Opel in Sports: Preparing for Serious Matters

"Motor sport should be people's sport!" This new slogan was meant to wipe out the old equation of "automobile" with "rich man." For this, cars were needed that, as "people's cars", could interest all levels of the population. Since 1924, Opel had been building such a car, the "Laubfrosch", a useful vehicle of the highest precision that could be made economically in large series. In 1932 the Adam Opel AG felt itself challenged to put the performance of their large-series vehicles to the test. In the very next year, as Germany's sporting events were officially reorganized, they took part in the two biggest and most important national competitions, the "Drei-Tage-Harz-Prüfung" and the first "2000 km durch Deutschland" run.

A series of other reliability trials followed every year, with not only factory teams taking part, but also Opel drivers from all over Germany: the "Winterbewerb Oberstaufen", "Internationale Alpenfahrt", "Deutschlandfahrt nach Garmisch-Partenkirchen", "6000-km-Italien-Rundfahrt", "Brandenburgische Geländefahrt" and the "Geländezuverlässigkeitsfahren" that were part of the "Deutsche Kampfspiele" in Nürnberg. These paramilitary tests of drivers and automobiles, conducted by the NSKK, the "Nationalsozialistische Kraftfahrerkorps", included steep mountain roads, gravel roads, driving through water, and off-road orientation tests.

Although they were not originally designed for these events, Opel cars proved themselves well in many sporting competitions. Only in 1936 did Opel develop special bodies for their six-cylinder cars to use in NSKK sporting events. These two-liter and later Super-6 sport two-seaters, as well as special off-road sports cars, exhibited the capability and reliability of the Opel designs and the parts they used in many sporting competitions.

Not at top speed – survival is what matters: An Opel 1.2-liter struggles through the sand on a steep grade during the 1934 Brandenburg Cross-Country Run.

Trucks also took part in these tests, and later became the backbone of the Wehrmacht. This two-ton 1934 Opel Blitz is seen in the Black Forest in 1935.

The Opel 6 roadster was made for such tests as this East Prussian Tour of 1935.

In these competitions the Reichswehr also tested the suitability of the Kübelwagen, here on the Opel 1.2-liter chassis.

A special body for an Opel 6 taking part in the 1935 Winter Trials.

Although Opel cars regularly proved their abilities in the winter and mountain trials, there was no specific development contract for Opel from the German Wehrmacht, which was newly reorganized since 1935. Neither a Kübelwagen nor a light uniform car was to be developed at Opel, for the new regime's mistrust of the American management in power at Rüsselsheim was too great. The motorization of more and more troop units, which the Reichswehr had begun in the late twenties, led, because of limited financial means, only to a few special military designs. The Reichswehr limited itself at first to stock vehicles, which took part in the various competitions. The Army's own passenger cars were usually fitted with light two-seater bodies by Trutz of Coburg, but there were also several other coachbuilding firms that converted production vehicles into open Kübelwagen. Later on, many Opel delivery trucks and sedans were rebuilt into Kübelwagen and light transport trucks.

In 1936 the Opel AG surprised the automotive world with a new Opel 6 off-road sports car, with a shorter wheelbase of 2390 mm and improved ground clearance of about 300 mm. After its first trials, it was eligible for the NSKK competitions and the many off-road trials. It came equipped with off-road tires, folding windshield, bottom boards, strengthened jack, spade, flasks, and a fuel tank enlarged to 110 liters. The engine compartment was also protected to keep out water at the bottom. These cars made their debut at the East Prussian Tour of 1936.

The Super-6 off-road sports cars also adorned the front page of "Deutsche Kraftfahrt" in June 1939. The magazine had merged with the "DDAC Motorwelt."

Whatever experience could be gained for military use in the sportingly camouflaged competitions was immediately utilized in series production. In off-road competition, the manufacturers used higher engine compressions, different gears and ratios. Better ground clearance was attained with strengthened springs and larger off-road tires. Despite the good experience with modified production vehicles, the military requested special designs, such as were defined in the so-called uniform vehicles as a basis for special bodies. In particular, the interchangeability of many parts that had to be replaced was to be attained through reduction to a small number of types.

The uniformity and practicality to be attained in the German automobile industry were called for by the general in charge of motor vehicles, Oberst i.G. Adolf von Schell, in March 1939. The requirements for reducing the number of types produced by the German auto industry and for achieving uniformity of aggregates within the motor vehicle program, known as the "Schell Plan", was already substantially fulfilled by Opel in its new models of 1938.

Uniformity within the Opel vehicle program. In addition to the similar versions of Kadett (0.5-ton) and Olympia (1-ton) passenger cars and delivery vans, the following were identical:

Engine	Super 6, Kapitän, 1.5-ton Blitz truck. Admiral and 3-ton Blitz truck.
Clutch	Olympia, 1-ton delivery truck, Super 6, 1.5-ton Blitz truck. Admiral and 3-ton Blitz truck.
Gearbox	Olympia, 0.5- and 1-ton delivery trucks, Super 6 and Admiral. The 1.5-ton Blitz truck was similar.
Rear Axle	Super 6 and 1-ton delivery truck. Admiral and 1.5-ton delivery truck.
Brakes	Kadett, Olympia, 1.5-ton delivery truck. Larger brake shoes: Super 6 and 1-ton delivery truck. Master cylinder and hand brake alike: Admiral and 1-ton Blitz truck.
Steering	Kadett, 0.5- and 1-ton delivery trucks, Olympia.
Doors	Kadett and Olympia. 1-ton delivery truck, 1.5- and 3-ton trucks.

Nothing was made easy for the cars: Two-liter Opel roadsters in the East Prussian Tour in late January 1935, and crossing a river during the 1935 Brandenburg Off-Road Trials.

East Prussian Tour, 1936: The new off-road sports cars proved themselves excellently. But the Wehrmacht developed its own "light off-road-capable motor vehicle."

Echo Continental

Auf Grund des Wehrleistungsgesetzes haben Sie........

Auch Fahrzeuge werden einberufen.

Es gibt bekannterweise vierbeinige Kameraden. Die wechselseitigen Beziehungen zwischen Mensch und Maschine schufen jedoch einen neuen Begriff — den des Kameraden, der auf vier Rädern läuft. Das muß so sein und deutet nur auf ein gutes Verhältnis des Menschen zur Technik hin. Gerade daran aber erkennt man auch die Wichtigkeit des Kraftwagens, der heute zu seinem Teil zum Siege beitragen muß. Nur deshalb ruft ihn sein Gestellungsbefehl, wie er auch den Soldaten ruft, führt man über ihn Karteikarten und schätzt ihn. Er soll Helfer sein, Kamerad des Soldaten, wenn er gebraucht wird. Besser — und das wird den Abschied erleichtern — ist es, den Weg des Erfolges zu ziehen, als nutzlos im Winkel zu rosten.

»Gestellungsbefehl« und »Wehrpaßnotiz«

Die Untersuchungskommission bespricht den Befund ....

The "Echo Continental" periodical reported on the "war service obligation" of private motor vehicles.

Now the family must say farewell to a "loyal member."

Mobilization: Opel in the War

Although no Opel vehicle was designed especially for the Wehrmacht, Opel vehicles and technology were encountered everywhere in military life. The majority were "requisitioned" vehicles, or those confiscated in occupied territories. Opel was not only the most often-used make of car in the German Reich, for over half of its production was exported. But it was not just this frequency of civilian use that resulted in a greater than average use of Opel vehicles in the military. Conventional rear-axle drive, an easily starting engine, and the often-praised reliability of all components helped the soldiers out of many a mudhole or across pathless terrain. Opel engines also endured the frosty nights in Russia without complaint, and could be used in the morning to start other vehicles.

In the thirties, Opel did not develop any special military vehicles. The prototype of a four-door tourer on the basis of the 1.3-liter company car did not lead to the hoped-for large-series contract of a simple militarily usable uniform vehicle for the police and firefighters. The military saw more advantage in the off-road-capable uniform car that was produced by firms like Stoewer, BMW and Hanomag. These vehicles, though, turned out to be too delicate for tough wartime use, and also too expensive. Only as of 1942 did the VW Kübelwagen fulfill the requirements of the military for a light off-road passenger vehicle.

At the beginning of the war, a few units used types of Kübelwagen that were rebodied on the basis of requisitioned 1.2-liter or P4 company cars. Thus the advantages of a robust chassis were kept and augmented by stronger springs and larger off-road tires. These vehicles seldom looked much like each other, since countless coachbuilding firms worked on them, using their own tools and rebuilding plans. Large numbers of the P4 Kübelwagen were turned over to the police or militias in occupied regions. The small Opel vehicles served not only as combat vehicles,

Even this Super-6 cabriolet, just a year old, was recruited for military service.

but were also used at bases in Germany or by the back-line units. Some were rebuilt as "siren vehicles" for mobile public warning use, officially designated "Kraftfahrsirenen." Production Opel box vans were used as loudspeaker and film trucks by the Reich Propaganda Office.

Opel sedans and convertibles served not only as command vehicles, but were also used particularly by war correspondents. These also included photographers hired by the Opel works, so that we can turn back today to an entire series of photo documentations. In particular, the convertible version of the "Kapitän", first put into production at the beginning of 1939, was used by commanders and staffs, while the 3.6-liter "Admiral" was driven by higher commanders and staffs. "Admiral" sedans were also rebuilt as light transporters, ambulances and towing vehicles for light Flak and Pak guns. To be sure, these cars, developed as large touring sedans, were overstressed in tough wartime use. In addition, the 3.6-liter engines consumed much fuel, and thus could no longer be used in some units toward the end of the war. On the other hand, the 6-cylinder, 3.6-liter unit found new use in the Blitz 3-ton truck, so that more transporting capacity could be achieved with the same machine. This interchangeability of complete aggregates and parts gained great popularity for Opel among the vehicle repair troops.

The Opel Geschäftswagen was also used by the Reich Propaganda Office as a loudspeaker van for filming.

On the chassis of the Opel Geschäftswagen, the Reich Air Travel Ministry had the "siren vehicle" built, to be used by the police to replace non-functioning alarms as of February 24, 1939.

As a suggestion for a light personnel car, an open tourer was built, based on the P4 Geschäftswagen. It never went into series production.

Coachbuilders took up the idea of the P4 Kübelwagen and equipped the Balkan militias with them. The Hungarian army driving school, a unit of which is seen here, had more than fifty such Opel cars.

The 1.2-liter delivery van was also modified into a Kübelwagen by various small coachbuilding firms, and given greater off-road capability with special rear wheels.

An Opel loudspeaker truck in the field.

The police also used requisitioned private cars, like this Opel 1.2-liter convertible sedan and this Opel Kadett.

Many troop units preferred Kübelwagen and had requisitioned vehicles rebuilt by coachbuilders. Various versions were created, like these, based on two Super 6 Opels and one two liter. The two-liter Kübelwagen. The two-liter was photographed during a sleet storm in Crimea.

Private Opel sedans were hastily painted in dull military colors at army motor pools.

Private cars prepared for the Luftwaffe wait for action at a garage. Among them is a one-ton Blitz bakery van.

The advance is hastened with requisitioned private cars from Holland. This picture shows a 1938 Olympia filled with soldiers crossing a pontoon bridge in May 1940.

In Austria too, Opels were requisitioned and marked with military symbols.

An old 1.3-liter Olympia, 1937 type, during a rest stop in the French campaign.

The Olympia convertible was particularly useful for leading columns.

The roads in Russia were often like washboards, but could be handled remarkably well by Opel's synchronized suspension.

Often the advance on Russia's muddy roads was possible only along railroad lines, as here in November 1941.

The small Opel Kadett was rarely found in the Wehrmacht. But the 1.1-liter car was sturdy and could, as seen here, be repaired easily with its tool kit.

This two-liter Opel of an NSKK transport brigade is fitted with a new leaf spring while being leaned against a building.

The Luftwaffe liked to use the Opel Kapitän, as here on the advance into France in the spring of 1940.

An Opel Kapitän, this one with an opening roof panel, on the advance.

Opel Super 6 and Kapitän in the field.

The Opel Kapitän on parade, and in a mudhole.

Reich Party Day 1938: For the diplomatic corps, the Adam Opel AG provided classy Admiral convertibles like these.

Admiral sedans, requisitioned from the Rhine Province, were also used by the Wehrmacht on important occasions.

The big Opel Admiral was granted only to the highest officers. Here a general and a colonel are on an inspection trip in the central sector of the eastern front.

The strong 3.6-liter engine and rugged chassis of the Admiral made a powerful light towing truck. Another advantage was that it used interchangeable Admiral and Blitz parts.

Medical aircraft brought wounded men to hospitals at home for special treatment. A Luftwaffe Admiral ambulance is picking up patients from a "white Ju."

Backbone of the Wehrmacht: The Opel Blitz in Many Versions

On account of the extensive fortifications of the western boundary of Germany, the Inspector General of German Roads, Dr.-Ing. Fritz Todt, called on the Reichskraftwagen-Betriebsverband (RKB) for cooperation in building the West Wall. Thereupon the RKB took over the organization of transport by the eleven to twelve thousand motor vehicles from private companies that were used in constructing the West Wall, beginning in 1938. Their experience with the most varied types of trucks was a big help to the military. The maneuverable and yet load-carrying Opel Blitz held an outstanding position as compared to its competitors of equal strength, so that the Wehrmacht quickly ordered large numbers.

With the proved box principle and replaceable parts used in more than one model, Opel vehicles and their technology were very popular in the Wehrmacht, which used many of them. More than 100,000 Opel Blitz trucks saw service with special functions, as supply and troop transporters, radio or repair trucks, ambulances, buses or towing vehicles. This was a number that no other factory attained during the war.

Since the monthly production of the German automobile industry at the beginning of the war year of 1940 was not sufficient, transport vehicles were recruited from the private sector for the Wehrmacht. Thus along with the Blitz truck types intended for military use, one could find many vehicles with low-side bodies or differently designed cabs in wartime photos. Above all, these included box vans from furniture movers, or Opel Blitz tour buses. While the requisitioned Blitz trucks were finished with special bodies in many cases, the buses generally remained unchanged. They merely sacrificed their chrome trim and colorful paint.

The few one-ton Opel trucks used by the military remained unchanged in terms of their bodies. They almost always served as courier vehicles or loudspeaker trucks for the Reich Propaganda Office. The light box vans were likewise used to deliver newspapers, magazines and mail to the troops. With a low inherent weight of 1340 kg, the smallest Blitz could carry a load of more than 1050 kg. For that reason, the one-ton Blitz, Type "2.0-12" and "5200", were used to tow the 2-cm Flak 30 gun, even after the war began.

The Type 5200 of the one-ton Blitz, built from 1938 to the beginning of 1940, had the renowned new 4-cylinder short-stroke engine of the Olympia 1.5 liter, producing 37 HP. As of the spring of 1938, a Blitz truck with a 1.5-ton payload was also marketed, using the new 2.5-liter, 55 HP engine of the "Super 6" of "Kapitän." Although this little 1.5-ton Blitz with the "2.5-32" type designation had good driving qualities and was also fitted for war service with longitudinal troop benches for military use, only 5000 of them were used by the military in the 1.5-ton class, which was not taken very seriously anyway. Its production had to be halted in mid-1942, since according to orders from the Prime Minister, Reich Marshal Göring, in charge of carrying out the four-year plan of March 2, 1939, the Schell Plan of type limitations had to be imposed on the motor vehicle industry. Thus in the 1.5-ton class, the S- and A- trucks with Steyr engines were produced uniformly by the firms of Steyr-Puch AG, Auto-Union AG and Phänomen-Werke Gustav Hiller AG.

The situation was different for the three-ton truck. Here the Opel design was to be adopted by the firms of Carl F.W. Borgward and Daimler-Benz AG.

Though the one-ton Opel was hardly prepared for field conditions when the war began, the lower photo shows the camouflaged headlights and other field equipment.

This was done, though, only by Mercedes, and only at a very late point in time. So the 3.6-liter three-ton Opel Blitz S, which was built only by Opel in Brandenburg from April 1937 to early August 1944 and at times also at the GM assembly plant in Copenhagen, has by far the highest numbers with 82,356 units built. The well-known 6-cylinder, 75 HP engine of the "Admiral" was later detuned to 68 HP for use in the Blitz to provide better torque for off-road use.

This reliable and nimble three-ton truck enjoyed great popularity among all sections of the troops. In the tumult of the war, the simple but sturdy Opel technology proved itself well and was far superior to many a special military design. What could not be repaired with tools carried on it could be handled easily, thanks to a comparatively good supply of spare parts, since the factories in Rüsselsheim and Brandenburg were producing plenty. The fact that Opel spare parts could be used interchangeably in several truck types was also a very important factor. Many supply sergeants wanted to have a motor pool as "purebred" as possible, with only Opel trucks, which they attained through all sorts of trades. Thus many a bottle of Cointreau was swapped to neighbor units for needed Opel spare parts during the French campaign. While Mercedes and BMW spare parts often had to be obtained by soldiers on furlough, such shortages of Opel parts occurred, according to a storekeeper sergeant, only in the case of radiator grilles, without which trucks could not be used.

Once one had used spirit lamps and hot water to get an Opel Blitz started after the coldest nights of the Russian campaign, it was that truck's honorable task to tow all the unit's other trucks and awaken them to new life. But it was not just for that reason that the Opel Blitz was preferred as a means of transport, especially in the Panzer regiments. Opel trucks all ran on gasoline, and there was more of this fuel available than Diesel fuel. The troops had also had bad experiences with the Uniform Diesel Truck. "When something was wrong with them, that was often the end." They were just as critical of the Mercedes three-ton Diesel, which was barely suited to the strains of the Russian campaign. Appropriate off-road and swamp testing had made this known even earlier. For in the tests, as later in the rough conditions of the Russian winter, the Opel Blitz, despite its two rear wheels, was superior to many other three-ton trucks thanks to its low weight. Where even the 4.5-ton Büssing-NAG s. gl. Lkw with its two driven axles failed, the Blitz truck still came through. When even the Blitz sank into a swamp, tracked vehicles had to help. For that reason, one expected much of the proclaimed Maultier halftrack, which was to be delivered to the troops soon.

For the "Propaganda-Blitz" there were important tasks on the front. Here a loudspeaker van is helped over a log bridge in May 1941.

This one-ton Blitz has had a rough time in the French campaign. In June 1940, German troops reached Montdidier, before the armistice agreement with Germany and Italy went into effect at 1:35 AM on June 25 of that year.

Captured "Tommys" on a one-ton Blitz in Belgium.

The 1.5-ton class was of little importance to the Wehrmacht. All the same, the Opel Blitz 1.5-ton was their most often-used light truck. For reasons of type limitation, the last 1.5-ton of 16,410 units produced was delivered to the Wehrmacht in November 1942.

Opel equipped the three-ton Blitz with various uniform bodies for the Wehrmacht. By the end of September 1940, 31,674 Blitz trucks had been delivered to the Wehrmacht.

Endless columns of supplies in Opel Blitz trucks rolled across the fields, as here in France on June 12, 1940.

Destroyed houses in the Belgian city of Tournai in May 1940.

Advancing in the West.

When the enemy blew up the bridges, makeshift ones were built, as here across the Meuse near Sedan on May 22, 1940.

It's 5:00 AM on September 30, 1939, and German troops in Opel Blitz trucks are crossing the Narev near Zambrow in Poland.

The wirelayers of the intelligence troops also liked to use the light and handy three-ton Blitz.

In Norway too, Opel truck were ready to being in captured Norwegian soldiers in April 1940.

Even special trucks once used by shipping firms were rebuilt for use as Wehrmacht transporters.

PK photos: Advancing in France with Opel Blitz three-ton trucks.

The roads in Russia did not allow a fast advance.

A muddy road on the Dniepr between Cherson and Berislav.

A radio station in a three-ton Opel Blitz at Fedorovka, west of Taganrog, before a typical Ukrainian farmhouse. Even in April 1942, straw mats were still used to keep motors warm, as well as for camouflage. To relieve the springs of the heavy body, especially when men entered the vehicle, the body was braced with rocks.

After the engine was warmed with fire, heat lamps or hot water, the truck was moved back and forth while in gear, to loosen the cold pistons and bearings.

Since the ferry over the Dniepr at Cherson could not run on account of floating ice, the column had to take a roundabout route 200 km long. The heavy staff buses in particular got stuck in the mud.

The advance into Russia demanded much of man and machine. This intelligence unit, with its heavy radio trailer, shows clearly what tasks could be entrusted to a production three-ton Opel.

In the streets of Simferopol in November 1941.

Before the Mariupol steel mills at the same time. In the foreground is a "little" 1.5-ton Opel Blitz.

For his Afrika-Korps, Rommel preferred Opel three-ton trucks. Legend has it that whole rows of Blitz trucks plodded through the desert, making clouds of dust to imitate advancing tanks and inspire the British to retreat.

On the way through the oasis of Driana near Benghazi.

Below: In Benghazi harbor, engineers built a loading ramp across a freighter resting on the bottom, on each side of which a ship could be unloaded. Many three-ton Opel Blitz trucks were sent to Africa.

North African theater of war: The water source at the oasis of Ummr-Rzem was already marked in various languages by September 1942. The Opel Blitz trucks were prepared for desert service with large single rear wheels.

This spare-parts depot consists of an Opel Blitz, only slightly protected from shrapnel by a low earthen wall.

A three-ton all-wheel-drive Blitz with uniform box body at the edge of Tripoli in September 1942.

From the coastal highlands, the Via Balbia swings down to the port city of Derna.

These Opel Blitz trucks carry supplies of ammunition and fuel.

Gasoline drums are being filled from tank cars.

The Opel Blitz three-ton "aircraft fuel tanker" (Kfz. 385). The open cab cannot have been a type of early uniform cab, but rather a special type suited to use on an airfield tanker, which obviously was based on the Mercedes LG 3000 (Kfz. 384). Vehicles of this type were also used as alcohol tankers (B-Stoffwagen) by the motorized rocket batteries (V2). Many Luftwaffe units, though, did not have special tankers, since they used normal three-ton Opel trucks with a tank and a pump at the rear.

A special tanker for hydrogen peroxide.

Unusual service for an Opel Blitz. The rear bed serves as a bier for a fallen comrade.

The end is near for a defective truck.

A shot-up Blitz truck.

The wreck of a burned-out truck in France.

Overturned, burned out and scrounged for parts . . . the end of many Opel trucks in World War II.

The Opel Maultier: Mobile Basis for Rocket Launchers

Under the designation "3.6-36 S/SSM", the Opel Blitz halftrack vehicle became a special version of the Blitz 3-ton-S truck. The halftrack "Maultier" (Mule) produced by Opel since October 1942 improved traction considerably in rough country. This truck type, promoted by the Waffen-SS, was to make forward progress possible, with a slightly reduced payload, in deep sand as well as on muddy, swampy ground. The "Gleisketten-Lastkraftwagen", as Sd.Kfz.3, was built by Klöckner-Humboldt-Deutz, Ford and Opel in the three-ton class and by Daimler-Benz in the 4.5-ton class. As of December 31, 1942, 1870 of them had been ordered, but by then only 635 had been delivered. Another 8500 were ordered by the Wehrmacht by May 1, 1943, and the German armaments industry was able to deliver more than 13,000 by the end of 1943, plus another 7,310 in 1944. More than 2100 trucks were rebuilt at the front. Ford delivered 13,952, Magirus some 2500 and Opel about 4000 of the Maultier, and they were all equipped with copies of the British Carden-Lloyd track running gear.

In the summer of 1942, Opel had developed a better, self-cleaning running gear unit, which could be mounted under the normal S Blitz with minor modifications. A few additional bore-holes sufficed to attach it, a shortened driveshaft drove the rear axle, which was moved forward, and the track wheels were simply screwed onto it. An additional braking system served for steering the manganese-hardcast tracks, which were identical to those of the Panzer I. While the Opel running gear allowed a top speed of only 38 kph, the Ford design, based on the Carden-Lloyd type, was said to reach a speed of 50 kph in top gear. The Wehrmacht, though, prescribed a top speed of 20 kph for the sake of making the tracks last longer. In Test Report IIId 4/43, the three different Maultier designs were compared.

In 1942 tests the Opel technicians demonstrated how easily a normal three-ton Blitz S could be turned into a powerful Maultier with the help of the running gear developed by Opel.

The chassis with cab and Opel running gear.

In comparison, the Opel Maultier with Carden-Lloyd running gear.

Comparative Values of the three Maultier Versions

Dimension	Maultier	Maultier	Maultier Opel	SS	LC (Ford)
Weight	for the	kg	6600	6600	6950
Payload	testing	kg	2900	2850	2900
Axle pressure	front loaded	kg	1000	1000	1420
Axle pressure	rear loaded	kg	5600	5600	4740
Length		meters	6.02	6.02	6.25
Width		meters	2.24	2.24	2.29
Height		meters	2.50	2.50	2.18
Ground clear.	rear	mm	320	320	275
Horsepower		HP	68	68	95
HP per ton		HP/ton	10.3	10.3	13.7
Fuel consump.	road	l/100 km	53	46.5	56.5
Fuel consump.	wood road l	/100 km	97	96	106
Top speed	low gear	kph	4.8	4.8	6
Top speed	top gear	kph	38	38	50
Track length		mm	1715	1800	1865
Ground pressure		kg/cm2	0.69	0.6	0.53
Steering brake			hydr	mech	mech
Brake control			foot	hand	hand
Turning circle with steer. brake		meters	13.2	13.3	16.0
Turning circle without steer. brake		meters	14.5	15.9	22.0
Track, front wheel to middle		meters	3.12	2.95	4.013

The Opel track running gear was not used for the production Maultier truck, although it was simpler to maintain and had better self-cleaning characteristics.

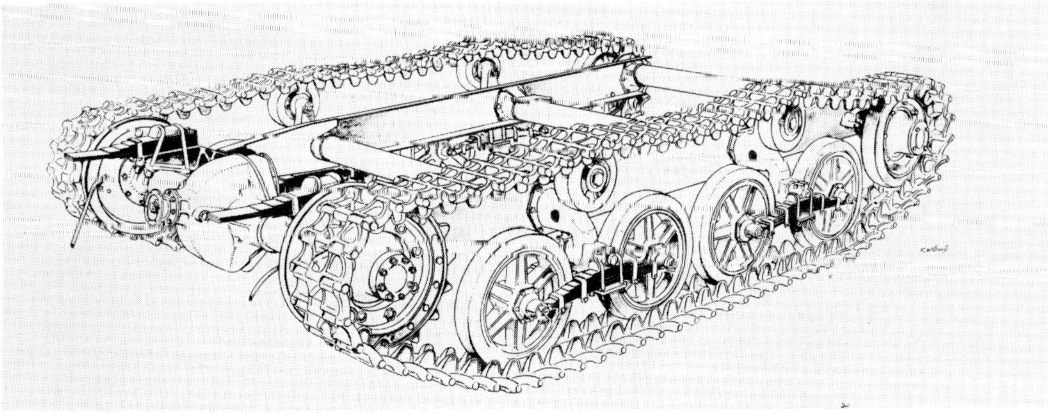

At the fronts, "Gleisketten-Lastkraftwagen" performed not only their usual supply work but were also given special tasks. In order to increase the mobility of small guns, they were mounted on the rear beds. Early in 1943 these experiences led to the development of an armored vehicle for the ten-barrel smoke-screen launcher. The Opel Maultier chassis was fitted with 8 mm armor and a 15 cm Panzerwerfer 43, swiveling 360 degrees, was mounted on it; this was designated Sd.Kfz. 4/1.

As of September 1943, 300 of these armored Blitz-Maultier vehicles were to be finished by the Karosseriewerken Westfalia, but only 90 of them were available as of August 1, 1943. In all, all of the ordered units were to be built, plus a small number of armored supply vehicles for ammunition transport. Today, only one restored Panzerwerfer on the Opel Maultier chassis can be found, in the Victory Memorial Museum in Belgium.

An Opel Maultier in front of a Messerschmitt Me 323 transport.

The uniform box body for cable-laying.

Armored Special Vehicle 4/1 on Opel Maultier chassis, produced by Westfalia.

A restored example at the Victory Memorial Museum in Belgium.

In December 1943, the few rocket launchers on self-propelled Blitz Maultier mounts were already in action. Despite the biting cold, there is much action on this runway near Newel, since the mud of earlier days has been hardened by frost. The column of "smoke-throwers" is led by an armored ammunition carrier.

WH-1408 820

An advertisement from "Motorschau", January 1941.

The All-Wheel-Drive Blitz: Ready for Service in the Roughest Terrain

The Wehrmacht's good experiences with the Opel Type S three-ton truck inspired the Commanding General of Motor Vehicles, Generalmajor von Schell, to urge Opel as early as 1938 to equip this vehicle with all-wheel drive in the simplest possible form for military purposes. With the lowest inherent weight and the greatest ruggedness of design, optimal off-road capability and utility for military purposes were to be attained. The development, which was begun at once, ultimately resulted in the concept of the 3-ton Allrad type, which now, instead of 34% (with normal rear-wheel drive), could climb a slope up to 70% with a full load thanks to its all-wheel drive.

The wheelbase of 3450 mm was shortened from that of the S type by 150 mm, moving the front axle closer to the cab. Thus the Blitz 3-ton "A" took on a bullish appearance. The Allrad-Blitz could also be distinguished easily from the S Blitz by its cut-off door angle and the engine-hood edge running straight over the fenders.

Technical changes to the A Type were limited to an additional gearbox and the front drive axle with differential. The separating gearbox could be shifted on or off for off-road driving with one movement of an extra shift lever, which was an important contribution to this vehicle's mobility. Since the off-road gear was set off in this gearbox, the five gears of the standard main drive were set off again, so that the driver of the all-wheel Blitz had in all ten forward and two reverse speeds at his disposal.

The all-wheel Blitz, in production since 1940, was first displayed to the public at the Vienna Spring Fair in 1941. Although over 35,000 all-wheel Blitz trucks – a number attained by no other all-wheel Wehrmacht truck – were built before the factory in Brandenburg were destroyed, the combat troops could not be supplied sufficiently with these desirable trucks. The production capacities for the intermediate gears and front-wheel parts were limited by the shortage of urgently needed raw materials. Thus at most five of these off-road-capable all-wheel Blitz trucks could be assigned to a motorized regiment. In a daily report of 1943, the use of the all-wheel 3-ton Blitz was already determined: Only urgently needed supply vehicles such as field kitchens, ambulances and radio trucks were to have all-wheel chassis. Some of the motorized radio unit bodies on all-wheel Blitz chassis were made in small numbers by the firm of Ludewig in Essen, which produced Wehrmacht omnibus bodies.

The driven front axle of the 3-ton-A.

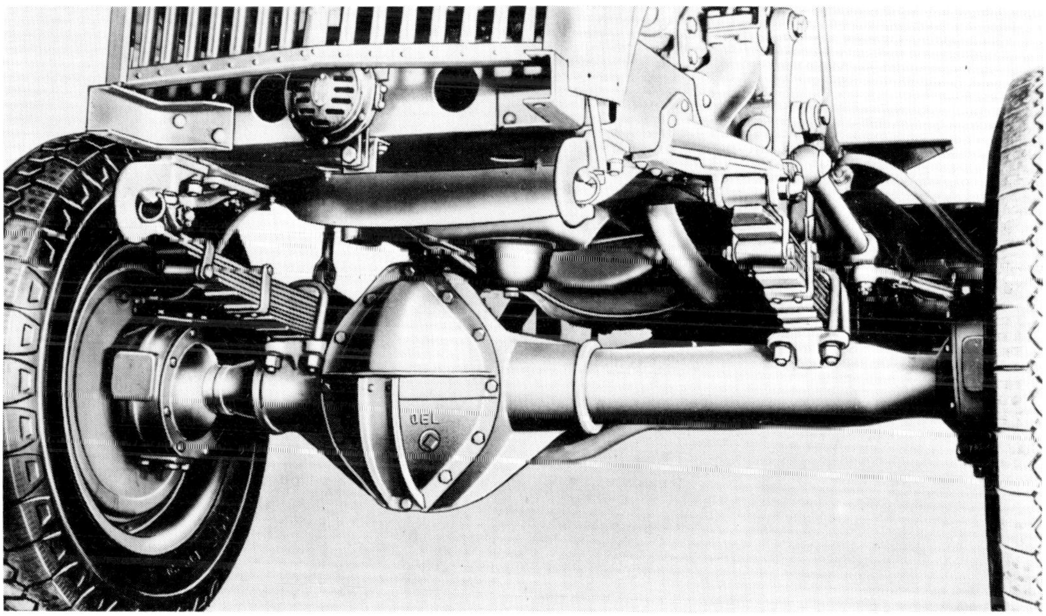

The chassis of the 1941 Blitz 3-ton Allrad.

The Blitz Allrad with uniform rear body.

The uniform box body on the Blitz 3-ton Allrad.

The uniform box body for spare parts supplies (EW234) in its material-saving 1943 form.

Test drives of Opel and Mercedes all-wheel-drive vehicles by the Wehrmacht in the winter of 1940-41.

A three-ton Allrad-Blitz that survived the war in Upper Bavaria and is now in the caring hands of a militaria collector. Likewise the NSU Kettenkrad with Opel Olympia engine.

A new body for a rescue truck on the chassis of an Allrad-Blitz, seen in the first postwar years.

The Opel Omnibus: Troop Transporter, Writing Room and Hospital

At the very beginning of Blitz production, there was, besides the two- and 2.5-ton trucks with 3400 and 4000 mm wheelbase, an extra-long 2.5-ton chassis 4650 mm long, which served as a sturdy chassis for buses holding 28 to 30 passengers. In 1938, when the new model was introduced, the Opel buses' share of the market increased from 25% in 1936 to 39%. In addition to the high cruising speed, this success was due primarily to the favorable ratio of the vehicle's weight to its payload. Built in the modern light construction style, the Blitz bus with 37 seats weighed 3500 kg and thus required only 95 kg of its weight per passenger. This then inspired many coachbuilding factories to equip low-frame Blitz bus chassis with streamlined aluminum bodies.

Field-gray paint for a once-bright "Autobahn-Blitz" bus on Opel chassis.

The Opel Blitz three-ton low-frame chassis for bus bodies, 1938 version.

Whether bus, workshop or delivery van body, there were many body types made by FKF, Kässbohrer, Kühn or Ludewig for the Blitz.

A Blitz with uniform bus body is loaded onto a ship in Naples, bound for Tripoli, in 1941.

But these Opel buses too, with their handsome "superhighway bodies", were taken from their proud owners for military use, since the forces lacked sufficient numbers of buses for fast troop transport. Neither intercity buses nor special tour buses survived their service in the Wehrmacht. Their handsome bodies were generally one-offs, sometimes built to suit the buyer. Only sightseeing buses with "convertible tops" were of no use to the military.

For city and intercity line use, bodies of conventional type were built, particularly for use with trailers, in which an Opel Blitz could be depended on to move up to ten tons. These vehicles usually were retained by the postal system and the railroad. In the last war years, newly created substitute vehicles were made, only with bodies including wood-gas burners, which were mounted at the rear of the bus.

Three Afrika Korps buses at a rest stop in the desert.

For lack of liquid fuel at home in 1944, wood-gas generators had to be installed in vehicles. Kässbohrer produced many generator buses on Opel three-ton chassis for the Deutsche Reichspost and Reichsbahn.

The "gas works" were concealed at the rear of the all-steel bus.

The Blitz buses with uniform bodies, built by the firm of Ludewig in Essen, saw Wehrmacht service in many ways: as troop transporters, mobile hospitals, writing rooms, or as command and intelligence offices. They were the simplified version of the usual production Blitz low-frame chassis with intercity bus bodies. The Ludewig Brothers delivered over 2880 of these buses with uniform bodies to the Wehrmacht between 1939 and 1944. While the body was made of sheet metal until 1943, the steel frames were fitted with composition panels as of 1944. At the same time, the interiors were modified through the use of wooden benches, which could also be used for carrying wounded men without taking up much space. These, as basic furnishings, were permanently attached to the vehicle and were part of the standard system of the 3-ton Wehrmacht ambulance. For use as a transporter of the wounded, the Blitz bus was also fitted with a heater that burned any fuel, which along with the variable furnishings reduced the capacity from 30 to 26 seats. As a mobile hospital, there was space for seven to ten badly wounded on stretchers and six or ten slightly injured, depending on circumstances.

These Wehrmacht buses had a radius of action of 280 kilometers and were ideal for fast troop transport on paved roads. As Model W39, these roomy vehicles served not only in the medical corps, but also as command or map cars, writing rooms, and intelligence centers. For these purposes, many modifications could be made with only a little work. Only special uses as mobile repair shops or "print-shop vehicles" and "compositors' vehicles for propaganda purposes" required appropriate interior furnishings. In part, these Opel low-frame chassis were fitted with bodies resembling the uniform box van body. Some of the uniform buses running in Germany continued to be used for passenger service after the war. Thus they also played an important part in peacetime rebuilding.

An Opel Blitz bus is the workroom of a division's command post. The vehicle is well camouflaged in the country. Earth wall and dugout not only afford ground-level entrance but also protection from shrapnel and enemy sight.

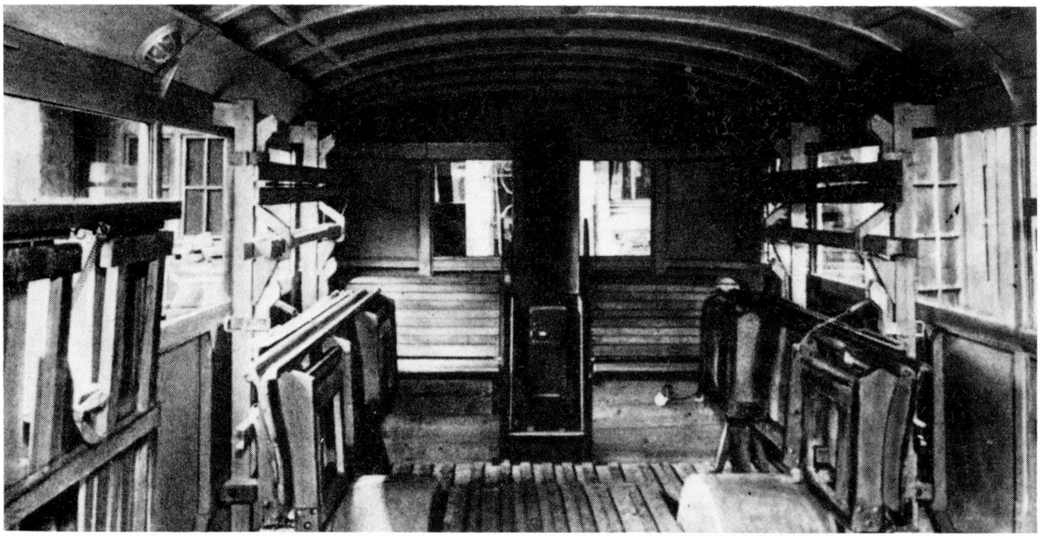

Inside a Wehrmacht bus finished in composition board as of 1944 and equipped with stretchers for wounded and a burning oven.

The Ludewig firm of Essen produced Wehrmacht uniform buses to the end. This last-series "Kom" was used as a mail bus after the war.

The Opel AG took over what remained of Blitz uniform truck production from Daimler-Benz. After the war, 467 3-ton Blitzes were built, over 100 of them low-frame chassis for new buses. This picture shows the factory bus fleet in 1950.

Opel Uniform Technology: Reliability in Every Piece

The proverbial reliability of Opel engines was responsible for the repeated recommendation of the Army Weapons Office, when the war began, that Opel engines be used for special vehicles. With the Opel "Super 6", a modern short-stroke engine, already in production in 1937, there followed in 1938 the 1.5-liter "Olympia" engine, a corresponding four-cylinder powerplant of the new type. These engines, using interchangeable parts, proved themselves splendidly in Opel vehicles, so that by 1940 the Army Weapons Office was already undertaking tests of a treaded motorcycle in 1940. It was driven by a 1.5-liter, 37 HP Opel engine, since the Ford Eifel engine recommended by its developer, Fritz von Falkenhain, could not be used. This towing vehicle, later renowned as the NSU Kettenkrad, was produced at Neckarsulm from 1941 to 1945, and later by Stoewer in Stettin, with 8536 units built. The treaded motorcycle (Sd.Kfz. 2) was described in the OKH memo of June 7, 1941 as a tractor to tow up to 450 kilograms. In the everyday events of the Russian campaign and the later "planned withdrawal", though, much heavier loads were towed by the small Opel engine. It was remarkable what these motors could accomplish in this small halftrack vehicle.

The nimble tracked cycle was equally desirable to soldiers and storekeeper sergeants. The Opel engine, its running speed slightly reduced, was familiar, robust and reliable. The installation of the water-cooled Olympia engine was done with the front face toward the rear in the center behind the driver's seat. The cooling system had to have a fan attachment with its drive fitted, so that the proper temperature could always be maintained, independently of the running speed. The top speed was stated as an unbelievable 70 kph, the maximum climbing ability 90%. A six-speed gearbox conducted the engine's power to the two track-drive wheels via a differential and lateral drives. The steering of the halftrack vehicle was done by turning the front wheel and by using the steering drive, which influenced the track running gear at any

The NSU Kettenkrad with four-cylinder Opel Olympia engine.

The driver's seat of the Kettenkrad.

turning of the front wheel beyond seven degrees. The track links were driven by forward-mounted drive wheels and carried the greatest part of the vehicle's weight.

The Kettenkrad with Opel engine proved itself so well that production was increased greatly:

1941 = 420 units
1942 = 985
1943 = 2450
1944 = 4490
1945 = 191

At least 531 of the Kettenkrad (according to serial numbers) were produced after the war, from 1946 on, and partially modified, also in France. They were used in agriculture and rebuilding. A planned large-scale continuation of production at NSU in Neckarsulm, though, did not take place.

Considerably smaller numbers of the 2.5-liter Opel engine were needed for another military vehicle: Somewhat more than 1000 examples of an amphibian car developed by designer Hanns Trippel and already displayed in September 1938. It had a 6-cylinder engine and was accepted by the Wehrmacht and SS in the period from 1941 to 1944. The Bank of the German Luftwaffe financed the designer's takeover

A Trippel S.G.6 of the last 1941 series, with six-cylinder Opel Kapitän engine, today owned by a collector.

A Trippel S.G.6 Schwimmwagen sedan during testing.

The "Biber" one-man U-boat had a four-cylinder Opel engine. A torpedo was slung on each side of the hull. The U-boat is now on display at the Technical Museum in Speyer.

of the occupied Bugatti works in Molsheim, Alsace to make series production of the Type S.G.-6" possible. The Trippelwagen with the Kapitän engine, with an operating weight of 1800 kg, reached 80 kph on the road and 12 kph in the water. With a ground clearance of 300 mm, a basinlike underbody and all-wheel drive with a five-speed special gearbox, the vehicle had fantastic off-road capability and could get onto land quickly, even with the worst shore conditions. The Trippel "S.G.-6" was usually used as a reconnaissance car, while a few with sedan bodies saw service with propaganda companies.

The reliable Opel engine in four-cylinder form was also utilized in another amphibian craft, which the German Navy had promoted in 1942 to support their elite troops. A small one-man submarine was to support swimmers on special missions and carry two torpedoes to near their target. This U-boat, called "Biber" (Beaver), had two powerplants: the Olympia engine, throttled down slightly to 36 HP, propelled the boat on the water and simultaneously charged the batteries to run the electric motor during dives. With a displacement of 6300 kg, a diving depth of some 15 meters could be attained. In 1943 and 1944, 325 "Biber" were built, of which only two examples are known today. The British captured one U-boat and set it up as a display in the Imperial War Museum in London. Another example was discovered while a channel was being dredged in Holland and turned over

After the French campaign, trucks were said to have been fitted with Opel Blitz engines at the Citroen works in Paris.

to the Technical Museum of Sinsheim and Speyer for restoration and display.

Equally "exotic" among Opel's war production was the use of the 3.6-liter engine in a transporter/towing tractor made by the Lauster firm of Stuttgart. The quarry owner Lauster developed an off-road tractor in 1942, with noteworthy pulling power and mobility in sand, mud and even swamps. Two further test models were built in 1943 as towing tractors or trucks. The tractor, designated "LW71", had two big rollers, each connected to an Opel engine detuned to 68 HP through a five-speed pre-selector gearbox and m.gl.Kfz.15, was also of interest. This vehicle, developed by the Army Weapons Office, was produced as of 1935 by Auto-Union at the Horch factory and powered by the 3.5-liter V8 engine. Since the production capacities in Zwickau were not sufficient, the Wanderer factory also produced chassis which were fitted with Opel 6-cylinder engines. In addition, as of 1940 Opel had to deliver not only 3.6-liter engines thus capable of delivering enormous pulling power. Despite the inherent weight of 12.5 tons, a ten-ton payload was possible. The one-off vehicle, 10.2 meters long and three meters high and wide, aroused the interest of the Luftwaffe, which awarded a contract for towing tractors for airfields. The end of the war interrupted the development, which had already included a diesel-electric drive for the rollers.

The use of Opel production capacities to build the medium uniform passenger vehicle, designated for the medium uniform passenger vehicle, but also make its production facilities in Brandenburg available in addition to the Blitz assembly lines. These deliveries for the medium uniform Pkw are not confirmed by any Opel statistics that survived the turmoil of the war. Thus the vague figures of several thousand chassis and some 6000 engines for the chassis made at the Wanderer works must remain only guesses.

The Kfz.15 was produced in Brandenburg as of 1940 as Opel Type mPI, and powered by the 75 HP Admiral engine. But as early as 1943, the production of the medium uniform passenger vehicle was halted, since these cars had proved to be too expensive to maintain and were to be replaced for wartime use by the VW Kübelwagen, which could be produced much more economically.

The medium uniform passenger car, Kfz. 15, with antitank gun during testing at the Horch works.

These pictures from the description and service directions for medium vehicles (with support axle) of 5/6/40 (D663/1) clearly show the robust 3.6-liter Opel engine.

Bild 35. Fahrgestell mit Triebwerkblock Opel

1. Anhängekupplung
2. Auspuffendrohr
3. Schwimmer für Kraftstoffuhr
4. Reservetank
5. Kraftstoffeinfüllschraube
6. Oberer Lenker
7. Federtöpfe
8. Nadelgelenkwelle
9. Hinterachsantrieb
10. Laufrad
11. Stoßdämpfer
12. Verbindungsschlauch für hydraul. Bremszylinder
13. Seilbremszug
14. Haupttank
15. Sammler
16. Kardanwelle (vom Zwischengetriebe zum Hinterachsantrieb)
17. Flansch für Reserverad
18. Reserverad
19. Schalthebel für Zwischengetriebe
20. Schalthebel für Schaltgetriebe
21. Handbremshebel
22. Lenkrad
23. Bremsoelleitung
24. Entlüftungsleitung für Zwischengetriebe
25. Zwischengetriebe
26. Bremswelle für Handbremse
27. Kardanwelle (vom Schalt- zum Zwischengetriebe)
28. Schaltgetriebe
29. Kardanwelle (vom Zwischengetriebe zum Vorderachsantrieb)
30. Luftfilter
31. Zündspule
32. Zündstrom-Verteiler
33. Kraftstoffpumpe
34. Ansaugrohr
35. Kühlwasseraustrittsrohr mit Anschluß für Thermometer
36. Verbindungsschlauchstück
37. Kühler
38. Windflügel
39. Zughaken
40. Verteilerstück für Bremsoelleitung

One day's production of medium uniform cars with 3.6-liter Opel engines, seen in Brandenburg ready for transport to the coachbuilders.

This medium uniform car is seen in a parade at Neustadt on the Weinstrasse in 1941.

Planning Without Future: The New 3-Ton Opel Blitz

Although Opel was never directly involved in the development of weapons systems and special military vehicles, the firm was still able to become Europe's greatest producer of automobiles in the thirties, even without military contracts. The Opel commercial vehicles, which were taken over by the military only much later, had contributed to this growth.

In the years of 1934 and 1935, Opel's commercial vehicles already amounted to between 10% and 15% of their entire production. By 1937 it had climbed to 19%. In 1939, with car production still running at capacity, it was already 24%. In the end, the truck percentage reached 73%, after production of Opel cars had to be halted in 1940. In the next few years, this fate also struck the other German automobile factories. At rüsselsheim by 1944, just fourteen sedans and convertibles were assembled as one-offs and prototypes. It was a time of military equipment production: aircraft parts, motors, trucks. On the average, 1800 Opel Blitz trucks left the factory in Brandenburg every month. There were 77,670 S-Type vehicles including the Maultier, Allrad and Omnibus chassis made between 1940 and the destruction of the factory in the summer of 1944. In 1939 it was still possible to use the assembly lines in the GM factories at Copenhagen to meet the needs of the war economy and produce Opel Blitz trucks with wood-gas engines. For lack of gasoline, some city bus lines also switched in part to other fuels. One of these lines was that of

A statistic chart from Opel information given to the Ministry of Armament and Ammunition.

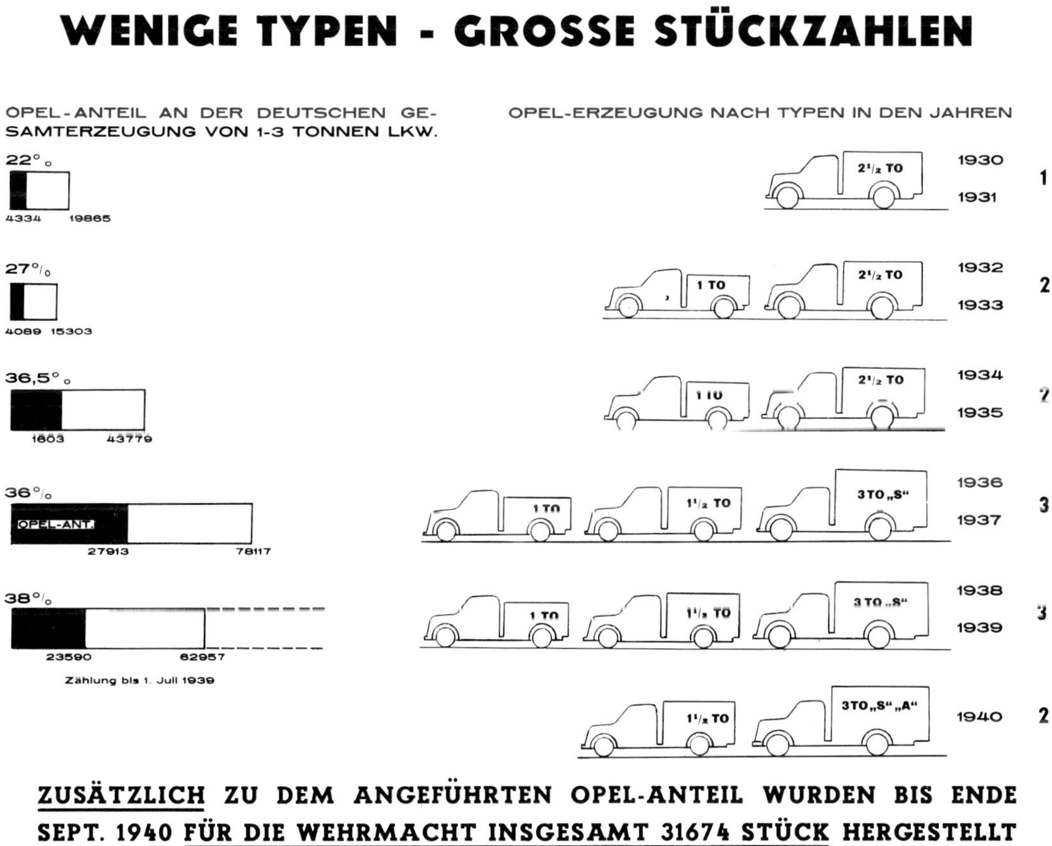

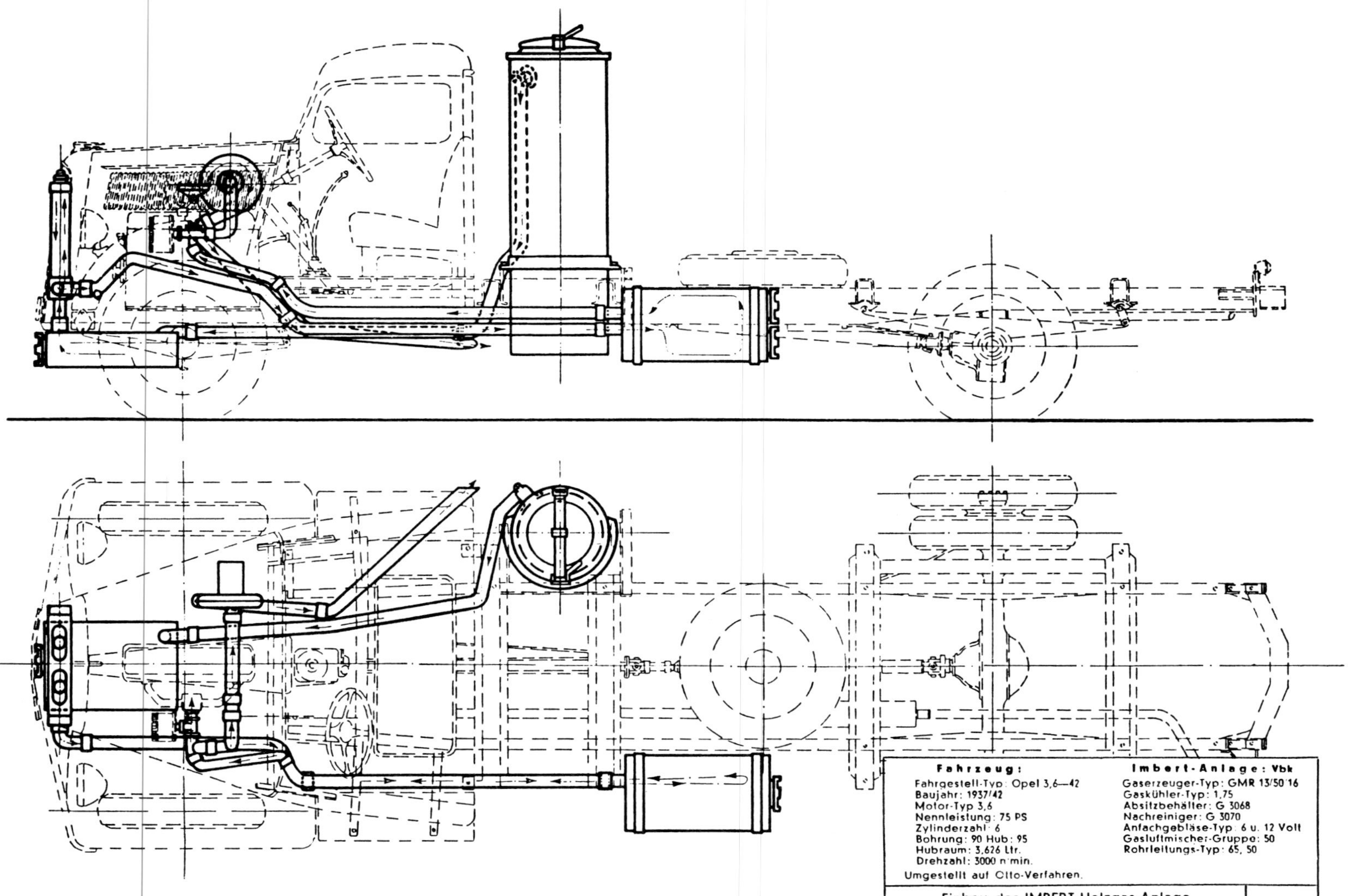

From the Opel factory handbook for wood-gas generator installation.

Before the factory doors in Brandenburg: As of 1943, vehicles had to be driven by wood-gas generators. Opel supplied three-ton Blitz trucks with Imbert wood-gas systems built in.

Dresden, which used Junkers R2 opposed-piston two-stroke Diesel engines to power Opel Blitz buses with KAFA bodies. For the postwar years, though, other engines were planned. During the war, Opel attempted to utilize the experienced gained with air-cooled Magirus engines for its own Blitz series. They developed an air-cooled 3.5-liter Opel engine with 72 HP at 3000 rpm. Three units were built and tested. Although the tests were completed successfully, the developments on the front lines prevented series production.

Tests made with new motors were only the natural continuation of the development of new trucks in the 3-ton Blitz series, begun in 1939. The experiments and prototypes that were only supposed to be modern in the fifties were already known in detail since 1941. Two basic concepts were worked on: The conventional long-hood design, but with a modern all-steel cab, and a more modern short-hood type, which caught the eye with its greater overall height. All-wheel-drive versions of both designs were planned. There was also an all-wheel-drive military version of the short-hood variety and with a canvas-covered cab, the Type 6600, created in 1942. Not even a photo of this cab-over-engine truck exists today in the archives of the Adam Opel AG. But in 1980 one could still see in Rüsselsheim a registered presentation document of November 1940, which showed the planned Opel model. Only a few technical modifications were made. A lighter, stronger rear axle and increasingly improved engine power with lower fuel consumption were desired. The changes in the body, though, were more obvious. The short-hood truck, with the same body length, had the advantage of a larger rear bed of 4000 mm, compared to the previous length of 3500 mm.

An Opel three-ton cab-over test truck with all-wheel drive.

When body development for the new three-ton Opel began, the American influence was clear. But new tools and production changes were impossible in the winter of 1940-41.

Allradantrieb und überbautes Führerhaus ergibt die gute Lösung für das Geländefahrzeug.

These new developments promised Opel not only successful use in all theaters of war, but also an extension of the firm's leading position on the export market in the coming peacetime years. But after the decisive winter combat of 1941-42, a transfer of production to the new Blitz trucks was no longer feasible. In the middle of 1944, the production facilities in Rüsselsheim and Brandenburg were subjected to repeated Allied bomb attacks. Large portions of the factory premises were destroyed. Rebuilding in the last months of the war met with delays. The plan for a third large factory in the eastern part of the Reich also remained unfulfilled. In April 1945, a new beginning appeared to be impossible. In was the first year since 1899 in which no Opel vehicles could be produced.

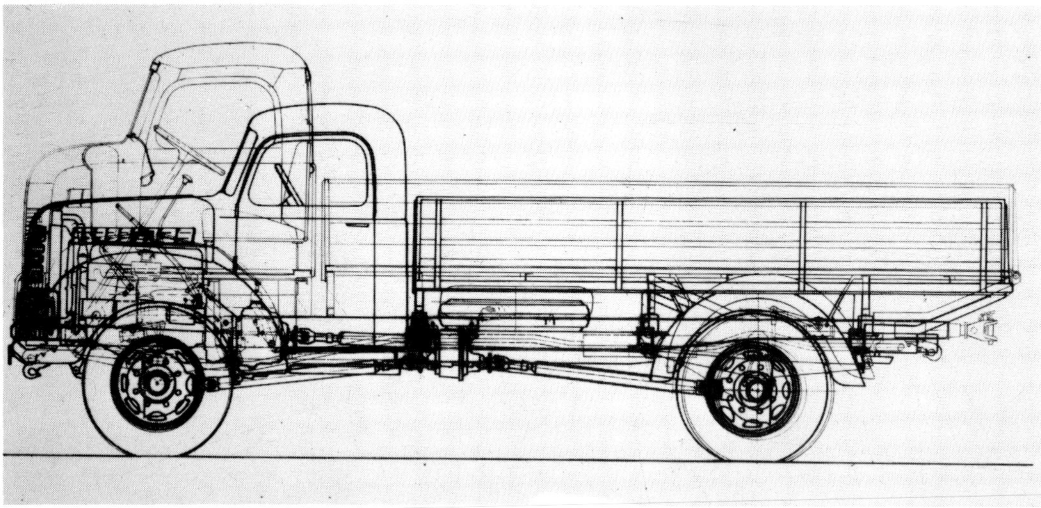

Drawing of the new Opel Blitz Type 6500 S.

The Brandenburg factory was a frequent target of air raids.

Without General Motors: The Works Amid Conflicting Interests

In 1929, the Adam Opel AG was 80% owned by the American auto manufacturer General Motors, and in 1931 GM purchased the remaining shares of stock. With that, the former family business was completely under American ownership. Against many national voices, the Opel AG had developed excellently since the GM strategists had influenced it. In almost all segments of the market, Opel had led the way during the course of the thirties. But Opel had not only become the market leader in passenger cars; the firm had also become Number One on the market in small trucks and reduced-tax three-ton trucks. Until the war broke out in 1939, Opel was seen as an important source of foreign currency for Germany on account of its large portion of exports.

Once the Polish campaign began, Opel's relationship with the United States entered a cooler phase. Even before December 11, 1941, the day Germany declared war on the United States, there were anti-American demonstrations before the gates of the Rüsselsheim works and regional repressive measures by the Gestapo against Opel employees. But the politically responsible parties could not agree about Opel as an "exemplary National Socialistic business": Through quarrels and conflicts of interest among the Luftwaffe Ministry, Nazi Party, Reich Economic Ministry, SS, and factory management, Opel remained without an obligatory enemy property administrator until the summer of 1942. Still in all, the top American executives had withdrawn from Rüsselsheim when the war broke out, and they no longer made use of their voice in the administration after December 1941.

Though the administrators installed by the Nazis were thoroughly successful in the sense of the party's goals in other firms, even with Ford of Cologne despite its foreign leadership, they did not want to risk any foreign domination of the two Opel production facilities in Rüsselsheim and Brandenburg. The Opel works in Brandenburg were doing a fine job producing trucks, engines, chassis and spare parts, working two shifts. In Rüsselsheim, the production of passenger cars, already considerably reduced, had been halted in November 1940, in order to make the entire facilities available for aircraft production. Since February 1940, Junkers aircraft engines and cockpits and canopies for the medium-range Ju 88 airplane, as well as running gear, frame components and complete electric networks with instrument panels, had been built there under license. As of December 1941, Opel had already reached a production volume of 200 million Reichsmark, and thus had become Germany's largest producer of motor vehicles. The deliveries for Junkers in Dessau. arranged by Göring, ended only at the beginning of 1945. The unsuccessful winter offensive from December 1941 to March 1942 resulted in the loss of over 31,000 trucks, which corresponded to half the previous year's production of the German auto industry. For the military, the capacities for ammunition, trucks and also aircraft urgently had to be raised. The experiences of series producers were called on, in order to transpose them to other branches of manufacturing. The capacities of factories that were running less efficiently were also to be reorganized on the basis of the "best businesses."

In the third year of the war, party members were not wanted in the German armaments industry as much as experienced efficiency experts. Even Daimler-Benz appointed a non-party member, Haspel, to be chairman of the board in mid-1942. The same clear signs were seen at the Opel AG: The technical director of the Rüsselsheim factory, Heinrich Wagner, was chosen to be the chairman of the board. At almost the same time, the disagreement as to who would manage the Opel enemy property was ended. Prof. Dr. Carl Luer was an "Opel Man" who had been a member of the board since 1935 and was still confirmed by GM as its chairman in 1941.

His influence as the administrator appointed by the Reich, though, remained little. Thus a "memorandum" for the Opel Blitz signed by him can be regarded as such. Luer and the new Inspector General of the Führer for Motor Vehicles, Jakob Werlin, newly appointed in January 1942, had agreed to triple the monthly production of trucks at the Brandenburg plant by using a third shift. Increased production of the Blitz trucks was certainly urgently needed, but for Brandenburg this would have been possible only through cooperation with the SS and recruitment of forced laborers and prisoners. Factory manager Heinz Nordhoff, though, was an experienced production chief and saw the effectivity of production endangered by the utilization of untrained and inept workers. Since the mid-thirties, and despite highly efficient production, Opel had employed a greater percentage of qualified employees than any other automobile factory.

Since the beginning of the war, no new funds had been available for the Brandenburg works, and in January 1942 Colonel von Schell had turned down a request to extend the factory. In the end, the directors in Rüsselsheim had also concluded, on May 10, 1942, that expansion at Brandenburg was not possible, since raw materials and qualified workers were already lacking for ongoing production. Not only was the technical chief Nordhoff against an expansion of production there, but the Ministry of Weapons and Ammunition, under the leadership of Albert Speer, also rejected expansion on the Havel, since the factory had too little protection against enemy bombers. Speer, who had expressed himself in principle in favor of the Opel Blitz as a medium-weight uniform truck for the Wehrmacht, wanted, after conferring with Hitler, to have an empty railroad-car factory in Riga expanded by Opel into an additional, self-financing assembly plant for trucks and spare parts. At the same time, production of 3.7 cm Flak guns was planned there. This procedure, though, was called off after inspection of the proposed production facilities. The plan to build a new factory and a central repair shop for Opel trucks and engines somewhere in the East, though, was kept alive.

Opel workers at Rüsselsheim assemble electric systems for the Ju88 built by Junkers in Dessau. Thanks to practical production methods, the transport of completed systems could be improved. The cable ducts could be folded and required only 1/3 the former space in railroad cars for shipping.

But the urgent need for Opel military trucks remained greater than the daily production. 50,000 trucks, more than 35% of the total three-ton truck production since 1939, had been delivered to the Wehrmacht by Opel in three years. Now other manufacturers, such as Borgward, Ford, Klöckner-Humboldt-Deutz and Daimler-Benz, were to make their production facilities ready to build the Opel three-ton uniform truck. The "Main Branch for Motor Vehicles" met at the Ministry of Weapons and Ammunition on June 22, 1942 with the task of distributing the required production of the Opel Blitz among the truck factories. Along with Borgward in Bremen, Daimler-Benz in Mannheim was given a contract to deliver 2000 to 2500 Opel trucks per month. Magirus in Ulm withdrew as a builder under license, since their truck production was to be given up. Ford of Cologne declared itself to be ready in principle to build its competitor's trucks under license, but decided in August 1942 to increase production of its own three-ton-S truck. Thus there remained only the Mercedes agreement and the request for Opel to build a new factory in the East to attain the quota of 9000 Opel-type uniform trucks per month expected by Hitler.

Women also worked on the assembly of electric systems.

Opel became the greatest supplier of aircraft components, especially for the Junkers Ju88 fighter.

The German three-ton uniform truck, Opel type, was also built by Daimler-Benz in Mannheim since early August 1944.

Last Request: The Blitz by Mercedes

The Opel Blitz had proved itself splendidly. All comparison and test drives had gone in its favor, and the demanding service at the front had shown over and over what qualities typified the Opel three-ton truck. Only a few improvements had been necessary for the S-Type. These included, starting with motor number Br 41-0538, a maintenance-free water pump, and as of 1942 an additional spring on the rear axle, to handle the heavy loads constantly carried at the front. Other modifications were required by the war, since increasing material shortages and delivery delays required simple solutions. Thus the last Opel cabs had a rear wall made of wood, in order to use less sheet metal. This demanded a noticeably higher roof shape, and a slightly steeper windshield slope. The wooden uniform cab required for other trucks, though, was not used on the original three-ton Opel Blitz. Only the late Blitz trucks built under license by Mercedes were fitted with this cab.

At the same time, all uniform trucks "of the war program, with regard to the required cooperative construction", were not to bear any maker's name or trade mark. But not only Mercedes-Benz had been foreseen as a cooperative builder of the dependable Opel Blitz by the Speer ministry. The Opel Blitz, recognized as a uniform truck far too late by the military, was to be produced by an "assembly circle" of Ford, Magirus and Borgward as well.

On June 25, 1942, Colonel von Schell, who had meanwhile lost his authority, signed the official building order. At this point it was clear that Magirus would halt production of trucks in favor of the Steyr-type Ost Bulldozer and Ford would, for the time being, continue to produce its three-ton S-Type. At Borgward there was no chance of new production because of bomb attacks. Only Opel and Daimler completed the licensing negotiations two months later. Opel received a one-time licensing fee of 800,000 Reichsmark. Later a compensation of 120 RM per vehicle was to be paid, specifically for the vehicles delivered to the Wehrmacht and Waffen-SS. The building contract was to end when the war ended and referred to construction of the S-Type with both wheelbase lengths (3.60 and 4.20 meters) as well as the Allrad-Blitz. At Mannheim, Daimler-Benz produced only the Blitz with 3.60-meter wheelbase, as preparations for construction took too long. Mercedes also received from Opel American machine tools and necessary forging equipment. In addition, many parts for assembly were to be delivered from Rüsselsheim by Opel: pipes and

A battery of Opel engines produced by Mercedes awaits installation, autumn 1944.

Chassis assembly at the Daimler-Benz Mannheim works.

Postwar problems: The uniform trucks built by Mercedes lack tires. The uniform cab was installed until production ended.

The earliest postwar Blitz. The emblem on the grille was not attached by Mercedes.

tubes, carburetors, air filters and fuel pumps ready for installation, as well as all the instruments and windshield-wiper mechanisms.

Despite precise conversion planning for a period of eleven months, the start of production was delayed another year, until August 1944. At the same time, Blitz production at Opel suddenly stopped. A number of British Liberator bombers, in an unusually precise daylight attack from an altitude of 8000 meters, destroyed large portions of the facilities in Brandenburg on August 6, 1944. Rebuilding could not be commenced until the war was over. In 1949 the Soviet Union demolished the Opel factory, valued at 22.1 million DM, as part of the agreed-on reparations plan.

Thus the Daimler-Benz AG became the only supplier of the urgently needed three-ton Opel trucks. The Mercedes-built Opel was designated Type L 701 and carried no maker's name. The license-built trucks were also recognized by the angular uniform cab and lack of hood trim. The angular Wehrmacht cab made of composition board was still built at Mannheim until 1948. Then Opel was able to provide the original cabs again, and was to receive Opel 3.6-liter, 73.5 HP engines in exchange. This did not take place, for despite two license extensions, the last L701 truck was finished at Mannheim on June 10, 1949. About 11,000 "Mercedes-Blitz" trucks were built in those five years, with the greatest numbers built in the rebuilding postwar years:

1945 = 747
1946 = 1497
1947 = 2001
1948 = 3805

The remaining machines were taken over by the Opel AG in Rüsselsheim and used to build another 467 Blitz trucks from 1950 to 1954, as well as low-frame chassis for buses, urgently needed for transportation. These were the last three-ton Opels, since the GM management wanted to limit production to the small 1.75-ton type with six-cylinder gasoline engines, while Mercedes had, since the end of the war, developed its new Type L312 Diesel truck, based on the Opel, and attained great sales success. This model immediately attracted the attention of the newly established German Border Patrol, and later that of the Bundeswehr as well.

The destroyed factory in Rüsselsheim in 1945.

The prototype of the new "Super-Olympia" for "peacetime" was also destroyed in the firestorm. The successor to the 1938 model was to have a trunk lid and steering-post shift lever.

Military Orders: New Beginning in Rüsselsheim

On March 22, 1945, US units captured Mainz, and two days later they reached the right bank of the Rhine at Oppenheim. With that, the new masters were not far from Rüsselsheim, which had already been evacuated by the Nazi big shots. In numerous bombing attacks, the Opel works had lost over 50% of their production capacity. In addition, the victorious powers, after the surrender on May 8, 1945, called for the dismantling of the still-remaining facilities. The dismantled machinery and the production equipment for the small Opel Kadett car were requested by the Soviets. In the agreed-on reparation payments, the complete production line for the four-door Kadett, including all drawings and equipment, such as casting forms and patterns, were loaded onto 59 freight cars, which left the destroyed railroad station of Rüsselsheim on June 14, 1946, headed eastward. In the summer of 1947, the little Opel turned up in the Soviet Union as the Moskvitch 400, and was at first available only to the Red Army. Only as of the autumn of 1950 could civilian customers, including those in the West, order a Kadett that was called Moskvitch. But the interest among Opel customers remained meager. In the days of the "Cold War", they did not want to remember that the Soviet forces had dismantled the Brandenburg works too, down to the last water faucet.

Rebuilding with Opel Blitz trucks.

The US Army governs the Rüsselsheim factory. Vehicles of all kinds were brought in for repairs. An Opel Admiral convertible, already in US service, is being unloaded. At left are Kapitän-39 cars, already driveable.

While the Soviets were the occupying powers in Brandenburg, the Americans were in charge in Rüsselsheim. In July 1945, the Commander of the 7th Army, General Geoffrey Keyes, instructed that all defective US military personnel cars were to be gathered on the premises of the former GM branch of Opel. In 1941, the US Army had ordered 15,000 standard sedans for military use in the USA, and also "blocked" another 9000 vehicles at dealerships in the States in case the USA entered the war. Some of these Ford, Chevrolet and Plymouth vehicles of the 1941-42 model year accompanied the fighting forces to Europe and were to be spruced up for better-class personnel transport after the end of the war. These cars were almost all four-door sedans, since only 18% of the orders were for two-door sedans, all of which remained in the States.

To repair military vehicles, the occupying forces made use of the remaining facilities of BMW, Fiat in Turin, and Daimler-Benz at the former Allgaier Works in Uhingen. But Opel of Rüsselsheim was also to repair US sedans as well as still usable German models.

Above: A 1938 Cadillac is checked by an ordnance officer after being repaired. Inside the hood is the registry number 16522. General Patton drove such a Cadillac, numbered 16417.

Before being repaired at Rüsselsheim, the US cars are disassembled, like this 1942 Plymouth P14. *[page 138]*

Removed engines were cleaned. In the background are various American cars: a Cadillac, a 1941 Ford Fordor and a 1940 Chevrolet.

Next the cars were put on an assembly line for body repairs, as is being done with this 1941 Plymouth P11.

This 1942 Plymouth is being repainted. During overhauling, the body was not always removed from the chassis.

"Final assembly line": 1940 and 1942 Chevrolets.

The Opel repair shop in 1945: a nice mixture of American cars such as the Plymouth Town Sedan, Chevrolet Special Deluxe 1942, and Cadillac Imperial sedan, one of 21 built in 1943. The Plymouth in front can be traced back to San Francisco by its 141769 registration number. It was one of 2759 "frozen" cars that the US Army took from dealers' stocks in California in 1943.

A test drive on the Opel test track in Rüsselsheim.

The Opel workers accepted this task with thanks. In the period from 1945 to May 1948, exactly 1516 German and American automobiles were repaired and delivered to the US forces as good as new. The repair work, of course, was not easy at that time, with its energy and material shortages. In many cases only trading could help. To be sure, the Opel works developed more quickly than expected into the most important repair facility for the US Army. Mainly GM makes such as Chevrolet and a few Pontiac station wagons and Cadillac six-seaters were sent there, but Fords and Plymouths also had to be repaired. The engines were removed, cleaned and reinstalled, bodies were straightened, parts replaced when needed, and painted. But many fancy German limousines of former Nazi bigwigs were also delivered by US officers for overhauling, and later shipped to the USA.

But the U.S. officers also lacked another product: In the USA, the refrigerator already ranked among standard kitchen appliances. In Germany, the supply of household refrigerators developed only at the beginning of the fifties. Via the newly revived relationship with General Motors, the refrigerator production that had been negotiated between Opel and Frigidaire as early as 1937 was resumed. To make the new start easier, the military turned over equipment and important raw materials for the first Frigidaire refrigerators that were produced on an assembly line in the Rüsselsheim works from 1946 on. In the first year, only 573 refrigerators were built, in 1947 there were 4000, and in the following years the numbers kept doubling.

British occupation troops also obtained Opel cars, like this 1949 Kapitän in Hamburg.

Despite the ban on contact, whoever could drive a big Opel car in the US Army could make an exception to the rule.

Used military vehicles were very popular among used-car dealers, as long as they could pep them up with a little effort, as there were often few kilometers on the odometer. In July 1957, 160 Opel Kapitän cars from US supplies wait at the border by Oeding, near Stadtlohn, for permission to be taken to Holland.

In the background of this development there was a constant dialogue between the US military authorities and the former owners of the Opel works, General motors. GM, which had written off its Opel shares on account of the war, originally wanted to take on the responsibility for their old premises only hesitantly. Only in February 1948 did a GM study group visit Rüsselsheim. They observed the already begun production of the small Opel Blitz and the four-cylinder Olympia car. The need for new cars in postwar Germany was great. Thus the correct decision was made: On November 1, 1948, the GM manager announced that he had once again taken on the leadership of the Opel works in Rüsselsheim. The production of the six-cylinder Opel Kapitän, prepared for since that summer, could begin at once. In just the first two months, the entire production of 213 Kapitän vehicles could be delivered to the US military authorities. In 1949, French and British occupying forces also obtained the desirable six-cylinder cars. One third of the production went to the military, this being 2424 sedans. Thus the occupying forces, with their need for new vehicles, gave the Opel works the necessary impulse to rebuild after the destruction of World War II.

As of 1952, the US Army obtained numerous Opel Blitz 1.75-ton ambulances in various versions. The first fourteen were completed by the Miesen coachbuilding firm.

The builder of the longer version is unknown.

Hamburg's first fire chief's car was an Opel 24/70 HP Phaeton of 1907.

Helpers in Need: Opel Fire Trucks and Ambulances

The automobile was discovered for military use only late. As noted elsewhere, along with the important tasks of moving supplies, other jobs such as the transport of personnel and wounded, and special duties such as firefighting and rescue, were assigned to the motor vehicle. Thus fire trucks and ambulances were quickly created on the basis of the Opel truck chassis. Changes were limited to softer suspension and different tires.

After a major fire at the Opel factory, fire pumps were developed, beginning in 1911, that were suitable for mounting on standard trucks. The truck motor powered a four-stage high-pressure centrifugal pump that could produce 15 atmospheres of pressure and 2000 liters per minute. These Opel fire trucks remained in service until the late twenties, when they were replaced by the new Blitz trucks with special firefighting equipment made by other manufacturers.

In 1938, firefighting too had to subordinate itself to military measures. Requirements for the building of vehicles came from the Reich Ministers of Air Travel and the Interior, who shared the task of providing protection from air raids. The formerly red fire trucks were painted dark green. As of 1939, the type limitations in the so-called Schell Plan applied, followed in 1940 by the reduction to three use classes for "firefighting vehicle building." Along with Mercedes and Magirus, Opel was also foreseen as one of the few chassis suppliers. Thus the Blitz one-ton chassis was replaced by the Kraftzugspritze KzS 8, based on the 1.5-ton Opel truck. Later, three-ton chassis were also used, known as of 1943 as the "LF 8." As a fire tanker, the Blitz three-ton in S or Allrad form was relatively seldom seen other than at military airfields, but these trucks gave excellent service to the Fire Police from 1943 on.

An Opel "Kapellenwagen" (Chapel Car) for the western armies in the spring of 1915. The rear could be opened to form an altar.

A three-ton Opel ambulance in World War I.

A Kühn ambulance on special Opel chassis, made in 1928.

An Opel fire truck, one of the Adam Opel firm's production models from 1912 to the mid-twenties.

Opel factory ambulances, circa 1926.

When the production of the 1.5-ton Opel Blitz could be resumed at Rüsselsheim under the supervision of the American military authorities, the first chassis for ambulance bodies were soon planned. Even before World War II, the firm of Christian Miesen in Bonn had successfully manufactured special bodies for ambulance and rescue use. Limited by the miserable supply situation of the first postwar years, ambulances were built on Blitz chassis, and Olympia and Kapitän vehicles were also rebuilt for patient transport, and later also for rescue use. To be sure, only the American and French armies could order such emergency vehicles. Only seldom did the Border Patrol or, later, the Bundeswehr order Opel vehicles for emergency use.

Because of the war situation, the first postwar Opel Blitz trucks were exported without the typical Blitz emblem, like this fire truck, seen in Canton Tessin, Switzerland.

In the presence of US officers, the first 1.5-ton Blitz comes off the temporarily repaired assembly line on July 15, 1946. The new postwar production began at this point. The few 1.5-ton Blitzes built of prewar parts by May 1946 with permission of the US forces were not counted as postwar products.

An Opel Blitz 1.5-ton ambulance.

An Opel Blitz 1.5-ton police personnel carrier.

Below: An Opel Blitz 1.5-ton KzS 8 used by the Vienna Fire Department.

An Opel Blitz 1.75-ton ambulance with Miesen body, owned by the Union Rheinische Braunkohlen Kraftstoff AG in Wesseling in 1958. Special bodies on production chassis were Opel's contribution to military use.

The TLF 15 was the first production fire tanker, built from 1939 on, using the three-ton Opel Blitz chassis. TLF cabs as of 1943 had a rear window of only about 80 mm diameter. In the city of Beeskow, between Berlin and Eisenhüttenstadt, there was a central gathering place for fire trucks during the war. These vehicles were almost exclusively under the Reich Air Ministry (RLM), with the exception of a few training vehicles. The truck shown here is a three-ton Blitz Type 6700 Allrad of May 1944, with front-mounted radiator, with body, including a 2500-liter tank, installed at Klöckner-Humboldt-Deutz at their Ulm factory.
These vehicles were used for quick firefighting intervention, particularly at airfields. 325 of them are thought to have been built.
Performance for the 1943 model was 1500 liters. The pump in the rear is a rotary type KHD PH 515, driven via driveshaft and branch shaft.
Maximum weight was 6800 kg, top speed was prescribed by the Wehrmacht as 35 kph.

The next three photos show the TLF Blitz, probably used for training at Beeskow, then taken by the US Army and turned over to "Feuermeldezentrale II" of the Salzburg Fire Department in 1946. At the end of the eighties, the truck passed into private hands. The pictures were taken in 1994 at the Hattingen Museum grounds.

A rare photo from above. The rear view was taken at the KHD delivery site in Ulm in 1943.

The Blitz uniform three-ton truck, made in Mannheim, as a fire truck, LF8, at KHD in Ulm, with composition-board body.

JMKw 259 fire truck on Blitz three-ton Allrad chassis, with uniform box body.

Two photos from the Koebe fire equipment factory in Luckenwalde: an open KOEBE body with front pump, the one above for the Neuendorf aircraft factory, the one below for the Junkers aircraft and motor works in Breslau.

A 2.5-ton Blitz of 1934 as a fire tanker for airfields.

A 2.5-ton Blitz bus chassis of 1934 as a fire truck (Koebe body), with about 1500 liters-per-minute capacity, for the fire department of the Adam Opel AG in Brandenburg on the Havel.

A KOEBE KzS 8 fire truck, built in 1937, on one-ton Blitz chassis, uniform fire truck of the German Luftschutz from 1936 to 1939, light body for one firefighting group, TS 8/8 pump and TSA trailer.

Opel Blitz 1.5-ton of 1938 fire crew truck, with body by Koebe of Luckenwalde, for the fire police of Seeburg in East Havelland.

The new cab of the 1.75-ton Opel Blitz of 1952. The seats for three persons can be moved 75 mm; behind them is the 68-liter fuel tank.

Three Little All-Wheel Blitzes: No Orders from the Bundeswehr

In 1952, the Adam Opel AG introduced a new Opel Blitz 1.75-ton type. At the same time, the "Deputy of the Federal Chancellor for questions involved with increasing the Allied troops" in the Chancellor's office was dealing with preparations for motorizing the Bundeswehr, which was soon to be established. Already existing NATO norms for motor vehicles were to be taken into consideration. At the same time, experience gained in World War II discouraged the development of special military and uniform vehicles, preferring developments of models in series production. So it was important that the German motor vehicle industry should prepare for the expected large orders from the young republic.

In a remarkably short time, the Federal chancellor's office was able to test the applicants for the individual vehicle groups. Three manufacturers wanted to build the Kübelwagen: Auto-Union, Borgward (Goliath) and Porsche. The decision, made in the spring of 1956, favored the DKW F91/4, Three firms also offered to build the small off-road-capable truck. Borgward was present again, but so were Hanomag and, for the first time, Opel. Since 1951 Borgward had been producing a light truck based on its B2000 and suitable, with all-wheel drive, for police, firefighting and Border Patrol use. The Border Patrol, in fact, already had its original supply of various Hanomag L28 all-wheel-drive trucks in use. Thus Opel had to try especially hard to compete with two experienced opponents.

The 1.75-ton 1952 Opel Blitz, at first producing 56 HP and with no rear fenders. As of 1955, 62 HP was stated.

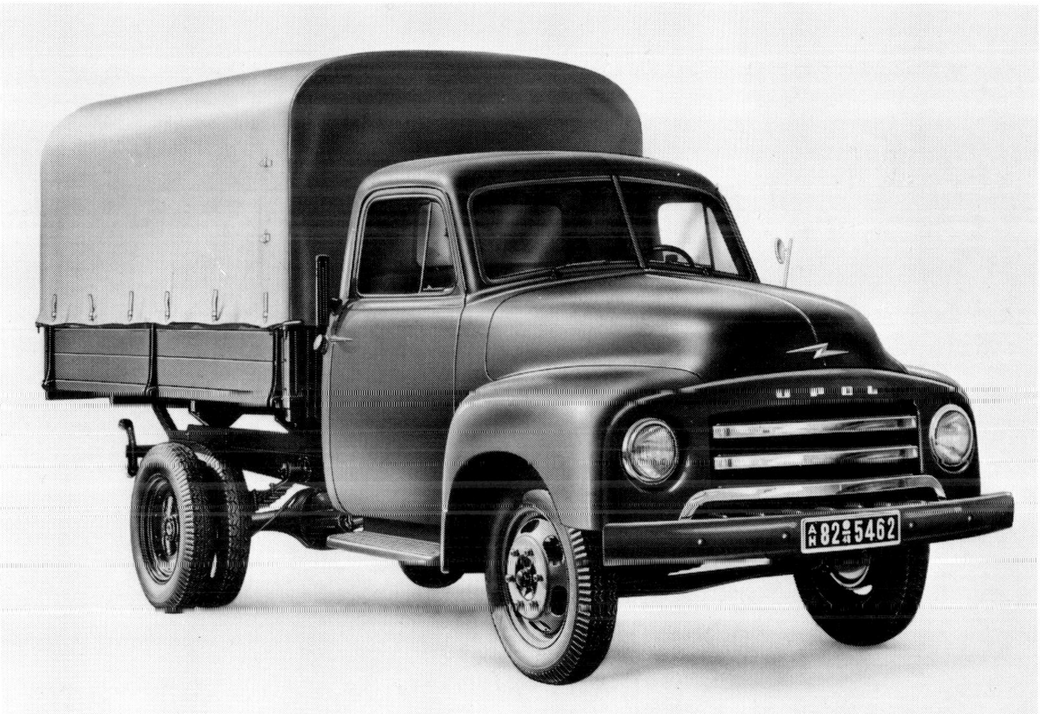

The development department built three all-wheel types in all, originally fitted with a slightly modified standard cab with a vertical radiator grille. In addition to a personnel carrier with high rear sides and canvas top, the firm of Miesen in Bonn provided a box body for radio trucks. The test vehicles, designated 1040/1 Spezialpritsche and 1040/2 Funkkoffer, were revised after a year and a half and, in October 1954, became the 1040/4 Kübelwagen made by the firm of Drauz and 1040/5, a normal open truck. By February 1956, both models had been driven more than 57,000 test kilometers, and contact had also been made with a third test truck, Model 1040/6, in reinforced form. The latter included a rear bumper that could be linked to the front bumper of the 1040/4 truck for towing purposes. With a pulling power of more than 4000 kg, this bent some 15 mm, and the front-wheel drive of the axle with Bendix-Weiss linkage broke. During backward towing with attachment to the front bumper and full torque in backward motion, several teeth of the crown gear broke and the large pinion bearing was completely destroyed.

In the winter of 1955, the Allrad-Blitz was first introduced to the "Deputy of the Federal Chancellor" along with other customers. The grille of the two models with rear bodies already had vertical bars.

But this experience was not all. The arrangement of the fuel tank in the passenger space of the Drauz bodied Kübelwagen was not acceptable. The brakes proved to be insufficient for off-road use and constantly prone to rusting from penetrating dirt. The exhaust system, which had been retained from the production model, also failed to stand up to off-road conditions. Likewise the steering support tube, which bent in the cab during off-road use and affected the steering. The problem was alleviated by the installation of a Hardy plate in the steering spindle. At the same time, it was recommended that the normal production steering wheel should not be approved for off-road use.

The two competitors in the 0.75-ton all-wheel class: Hanomag AL 28 and Borgward B2000 Allrad, both used successfully by the Border Patrol.

The Opel Blitz Allrad open truck and Drauz Kübelwagen during off-road test drives in the Hunsrück.

The Opel Blitz 0.75-ton Allrad first version, as an open truck and radio van, the latter with Miesen body.

The Blitz 0.75-ton Kübelwagen being tested according to NATO requirements.

And again, test drives for off-road capability.

A continual problem during the off-road test drives was the track rod, since it strongly influenced the ground clearance. Thus moving the track rod to a position ahead of the front axle was tried, as was dividing the track rod by an intermediate lever mounted on the front axle housing. The wading ability of both all-wheel-drive Blitzes was also unsatisfactory, as the engine stalled while driving through water. The tests showed that the radiator ventilator let moisture into the ignition system, and spark cables broke after just a few thousand kilometers and let water in.

The vehicles were tested until February 29, 1956 in the country in the northern Sonnwald in the Hunsrück area, and later in the Neupfalz area, near Stromberg. Every prototype began its test drives on the premises at Rüsselsheim and taken back there every day. This gave test results on the basis of 2/3 normal roads and 1/3 rough country, with fuel consumption of 24.5 liters per 100 km. The normal consumption for road use was stated as 16 liters per 100 km. The two all-wheel Blitzes had Phase II test engines numbered 54/V-0002L and 54 V/0003L installed. These were six-cylinder engines with a throttled-down performance of 62 HP at 3700 rpm and a compression of 6.5:1.

A problem to the end: the brake drums cannot stand mud and dirt.

Driving through a forest brook in the Neupfalz.

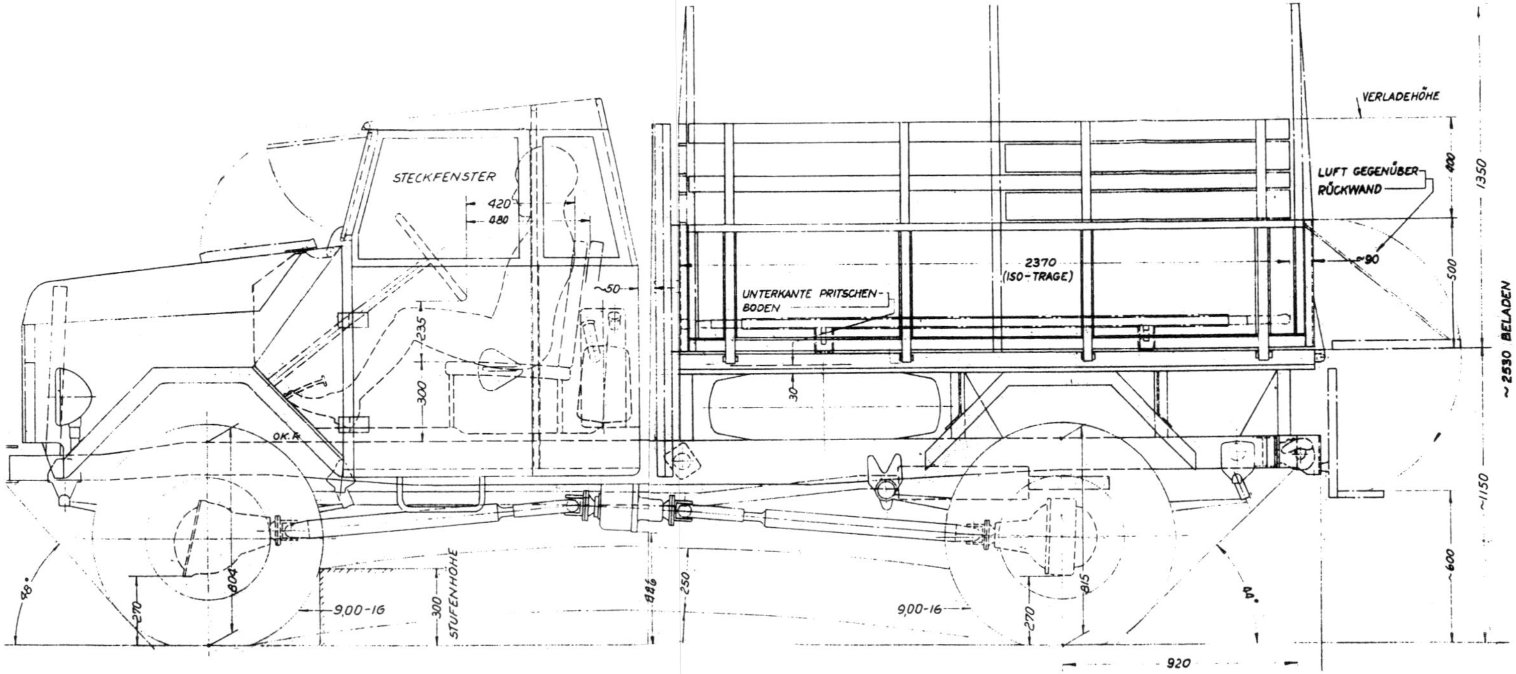

Drawing of the 0.75-ton Allrad Blitz truck, built to NATO standards.

The tests were not easy. No wonder that the engine picked up moisture and the vehicle stalled.

In January 1956 the Adam Opel AG presented its offer for three delivery versions. The new off-road A-0.75-ton Opel Blitz met the Defense Ministry's requirements then in force for tactical motor vehicles. One delivery was to be only the chassis for all kinds of special bodies, the chassis without wheel wells, with open cab and steel rear body with canvas top, and a chassis with box body for a nine-man crew. To be sure, the prices for the off-road Blitz vehicles were too high compared to the prices of the Borgward Group. For about 15,000 DM per 0.75-ton off-road vehicle, Borgward could supply more than 6700 vehicles in seven years, until 1961. They were in use as command vehicles, telephone-line trucks, and as trucks for repair and service units. The Borgward B 2000 A was also gladly used by the Bundeswehr as a driving-school vehicle.

At Opel, they consoled themselves over the lost contract by devoting more energy to expanding the production capacities for cars. The Bundeswehr and Border Patrol also bought their needed personnel cars from the Official Sales branch of the Adam Opel AG via the procurement offices. These Opel Rekord sedans, generally standard production cars, painted olive green, were available to staff officers as staff cars. In comparison to civilian customer sales, sales to the Bundeswehr were meager. But without the first contracts from the occupying authorities, production would not have come to new life from the ruins as quickly as it did.

The winners in the NATO competition: Borgward B2000 trucks and Kübelwagen.

Technical Data: Allrad-Blitz A-0.75-ton 1956

The off-road A-0.75-ton OPEL BLITZ corresponds to the established requirements of the Defense Ministry for tactical motor vehicles.

Planned Models

Chassis
Chassis with open cab and steel rear with canvas cover.
(without wheel wells)
Chassis with box body for nine men including driver.

Special Bodies

The chassis is suitable for all kinds of special bodies, as long as they will not exceed the prescribed maximum axle pressure on the chassis drawings, as well as the weights and measurements. It must be kept in mind that the available axle pressure of the loaded truck must not exceed the allowable gross weight of 3300 kg.

Engine

Type Opel 2.5 liter
Function four-stroke carburetor
Number of cylinders 6
Bore & stroke 80 x 82 mm
Displacement 2473 cc
Compression ratio 6.5 : 1
Performance at n = 3700 rpm 62 HP
Torque at n = 1200 rpm 15.6 mkg

Wheelbase 3300 mm

Weights only estimated

Max. allowable gross weight 3300 kg
Max. allowable front axle pressure 1600 kg
Max. allowable rear axle pressure 2200 kg
Chassis inherent weight ca. 1500 kg
Chassis carrying ability ca. 1800 kg
Weight in use as open truck ca. 2350 kg
Payload for open truck ca. 850 kg

Main dimensions of bodies and vehicles

Maximum body length from front angle of front wall to rear wall of body ca. 4150 mm
Max. allowable body width including attachments ca. 2000 mm
Max. allowable distance from front bumper to middle of front axle ca. 800 mm
Max. allowable rear overhang for body ca. 1030 mm
Limited by off-road angle required by Defense Ministry.

Kübelwagen

Overall length ca. 5120 mm
Overall width ca. 1900 mm
Hight over cover, unladen, ca. 2200 mm
Min. height for loading, ca. 1800 mm
Loading space behind driver's seat:
a) Main loading area length, ca. 1500 mm
Width ca. 1720 mm
Height ca. 1450 mm
b) Luggage space: length ca. 1100 mm
Width ca. 1720 mm
Height ca. 450 mm

Open Truck

Overall length ca. 5130 mm
Overall width ca. 1900 mm
Height over cover ca. 2600 mm
Min. height for loading ca. 1750 mm
Rear body length ca. 2500 mm
Rear body width ca. 1850 mm
Rear body wall height ca. 500 mm
Height of added wall portions ca. 350 mm

Ground Clearance and Off-Road Angle for laden vehicles

Front axle ground clearance ca. 270 mm
Rear axle ground clearance ca. 270 mm
Bottom clearance ca. 450 mm
Front off-road angle ca. 45 degrees
Rear off-road angle ca. 40 degrees

Turning Circle diameter ca. 14.70 m
Track Circle diameter ca. 13.75 m

Performance

Top speed on road with road gears ca. 95 kph
Lowest sustained speed off-road 3 kph
Wading depth to lower edge of frame in quiet water ca. 650 mm
Max. climbing ability in 1st road gear ca. 32 degrees
Max. climbing ability in 1st off-road gear ca. 85 degrees
Max. towing ability in 1st road gear ca. 1150 kg
Max. towing ability in 1st off-road gear ca. 2200 kg
Range on road with 20-liter reserve canister ca. 500 km
Normal fuel consumption on road ca. 16.0 liters per 100 km

A police team transporter of the Opel Blitz 1.9- or 2-ton series, built 1960-1965.

Production Figures 1904-1975
Opel Blitz, Delivery Vans and Trucks

Calendar	Blitz Vans & Trucks	Yearly Supply	Cumu-lative	Calendar	Blitz Vans & Trucks	Yearly Supply	Totals
1904		8	8	1940	19662	20172	182800
1905		17	25	1941	19432	19432	202232
1906		26	51	1942	8866	18866	221098
1907		24	75	1943	23232	23232	244330
1908		30	105	1944	16146	16146	260476
1909		47	152	1945	–	-	260476
1910		59	211	1946	839	839	261315
1911		112	323	1947	3219	3219	264534
1912		164	487	1948	7063	7063	271597
1913		151	638	1949	11574	11574	283171
1914		184	822	1950	7753	12756	295927
1915		1117	1939	1951	007	15055	310982
1916		1090	3029	1952	10424	21201	332183
1917		1134	4163	1953	9776	16675	348858
1918		1112	5275	1954	13165	19408	368266
1919		463	5738	1955	12235	18132	386398
1920		572	6310	1956	12642	18559	404957
1921		316	6626	1957	11375	16885	421842
1922		405	7031	1958	9602	19601	441443
1923		186	7217	1959	9505	19631	461074
1924		154	7371	1960	5915	19122	480196
1925		1186	8557	1961	10611	24446	504642
1926		1730	10287	1962	8313	19806	524448
1927		4916	15203	1963	8258	14916	539364
1928		7148	22351	1964	7637	17061	556425
1929		5111	27462	1965	5539	15474	571899
1930		2835	30297	1966	7722	12972	584871
1931	2716	4064	34361	1967	3540	9075	593946
1932	1792	3086	37447	1968	4070	8685	602631
1933	2298	3489	40936	1969	7202	12994	615625
1934	6988	8480	49416	1970	6461	9212	624837
1935	9511	12969	62385	1971	5201	6846	631683
1936	14260	21756	84141	1972	4514	117	637800
1937	17337	24174	108315	1973	4398	6173	643973
1938	21797	26377	134692	1974	3070	5381	649354
1939	24486	27936	162628	1975	58	1662	656397
				Total:	417201		

Opel Blitz Statistics

By Building Year

Model	1.1 liter 1.3 liter Geschäftswagen	1 ton	1.75 ton	2.5 ton	Total
1931	1.348	–	677	2.039	4.064
1932	1.294	–	–	1.792	3.086
1933	1.191	62	–	2.236	3.489
1934	1.492	4.038	–	2.950	8.480
1935	3.458	4.289	–	5.222	12.969
1936	7.496	5.427	–	8.833	21.756
1937	6.837	6.446	–	10.891	24.174
1938	4.499	1.361	5.577	14.859	26.296
1939	3.335	2.935	5.066	16.485	27.821
1940	410	379	1.678	17.605	20.072
1941	–	–	3.485	15.947	19.432
1942	–	–	604	18.262	18.866
1943	–	–	–	23.232	23.232
1944	–	–	–	16.146	16.146
1945	–	–	–	–	–

By Production Series

Model	Type	When Built	Quantity
P4 1.3 ltr Geschäftswagen	1396	11/1935 - 2/1938	15.672
P4 1.1 ltr Geschäftswagen	6100	3/1938 - 4/1940	7.405
Blitz 1 ton 6 cylinder	2.0-12	12/1933 - 3/1938	21.437
Blitz 1 ton 4 cylinder	5200	10/1938 - 6/1940	3.500
Blitz 1.5 ton	2.5-32	1/1938 - 11/1942	16.410
Blitz 2 ton/2.5 ton 3.5 ltr	3.5-34/83	12/1930 - 3/1937	21.196
Blitz 3 ton 3.5 ltr	3.5-36+47	5/1936 - 7/1937	5.605
Blitz 3 ton + 3 ton "S"*	3.6-36	4/1937 - 7/1944	82.356
Blitz 3 ton long	3.6-42	6/1937 - 7/1944	14.122
Blitz 3 ton low frame	3.6-47	5/1937 - 7/1944	8.336
Blitz 3 ton all-wheel	3.6-67	7/1940 - 7/1944	24.981
Blitz 3 ton/	3.6 ltr total	4/1937 - 8/1944	129.795

* includes Blitz 3 ton S/SSM Maultier" halftrack, ca. 4.000
and Panzerwerfer on Blitz S chassis.

Technical Data, OPEL BLITZ 1931 to 1945

Typ	1,5—29		2,0—12		2,5—32	3,5—34 2 t	3,5—57 2 t	3,5—34 2 t	3,5—57 2 t	3,5-57 2½ t
Baujahr	1938—40		1934—38		1938—42	1931—33	1931—33	1934—37	1934—37	1931—33
Motor										
Zylinderzahl	4		6		6	6	6	6	6	6
Bohrung . . . . . mm	80		67,5		80	79,38	79,38	79,38	79,38	79,38
Hub . . . . mm	74		90		82	117,48	117,48	117,48	117,48	117,48
Hubvolumen . . . . cm³	1488		1920		2473	3417	3417	3417	3417	3417
Bremsleistung . . . PS	37		36		55	64	64	64	64	64
Verdichtungsgrad . . .	6		5,75		6	5,2	5,2	5,2	5,2	5,2
Wechselgetriebe										
1. Gang	4,32		3,91		4,48	5,56	5,56	5,56	5,56	5,56
2. Gang	2,47		2,47		2,61	3,70	3,70	3,70	3,70	3,70
3. Gang	1,49		1,49		1,68	1,80	1,80	1,80	1,80	1,80
4. Gang	1		1		1	1	1	1	1	1
5. Gang	—		—		—	—	—	—	—	—
Rückwärtsgang	4,69		4,21		4,75	6,50	6,50	6,50	6,50	6,50
Verteilergetriebe										
Straßengang	—		—		—	—	—	—	—	—
Geländegang	—		—		—	—	—	—	—	—
Hinterachse										
Übersetzungsverhältnis	37:6		43:7		33:6	35:6 [2]	35:6 [2]	35:6 [2]	35:6 [2]	35:6 [2]
Übersetzungsverhältnis	—		—		—	34:5 [4]	34:5 [4]	34:5 [4]	34:5 [4]	34:5 [4]
Übersetzungsverhältnis	—		—		—	40:7 [3]	40:7 [3]	40:7 [3]	40:7 [3]	40:7 [3]
Übersetzungsverhältnis	—		—		—	41:6 [5]	41:6 [5]	41:6 [5]	41:6 [5]	41:6 [5]
Übersetzungsverhältnis	—		—		—	41:8 [6]	41:8 [6]	41:8 [6]	41:8 [6]	41:8 [6]
Räder und Reifen										
Reifengröße . . . vorn	5,50—18		5,50—17		5,50—18	6,00—20	6,00—20	6,00—20	6,00—20	6,00—20
Reifengröße . . . hinten	6,50—17		7,00—17		5,50—18	6,00—20	6,50—20	6,00—20	6,50—20	6,50—20
Felgengröße . . . vorn	3,25 E x 18		3,25 E x 17		3,25 E x 18	5″—20	5″—20	5″—20	5″—20	5″—20
Felgengröße . . . hinten	4,00 F x 17		4,00 F x 17		3,25 E x 18	5″—20	5″—20	5″—20	5″—20	5″—20
Reifendruck . . . vorn	2,75		2,50		2,75	4,25	4,25	4,25	4,25	4,25
Reifendruck . . . hinten	3,00		3,00		3,00	4,75	5,00	4,75	5,00	5,00
Elektr. Ausrüstung										
Batterie-Spannung Volt	6		6		6	6	6	6	6	6
Kapazität . . . . . Ah	75		72		75	90	90	90	90	90
Fahrgestell										
Radstand . . . . . mm	2900		2851		3250	3410	4000	3410	4000	4000
Spurweite vorn mm	1342		1346		1340	1433	1433	1433	1433	1433
Spurweite hinten mm	1400		1422		1360	1550	1550	1620 (1524)[1]	1620 (1524)[1]	1555
Wendekreis-Durchm. m	12,6		12,3		12,8	14,4	16,5	14,4	16,5	16,5
Wagenabmessungen	Pritsche	Kasten	Pritsche	Kasten						
Gr. Länge über alles mm	4700	4625	4635	4455	5450	5380	6370	5380	6380	6370
Gr. Breite über alles mm	1940	1680	1995	1705	1940	2130	2291	2291	2291	2130
Gr. Höhe über alles mm	2180	1865	2280	2060	2285	2428	2439	2580	2580	2439
Ladefläche	Pritsche	Kasten	Pritsche	Kasten						
Ladelänge . . . . . mm	2200	2150	2200	2115	2900	2500	3500	2500	3500	3500
Ladebreite . . . . mm	1800	1470	1800	1460	1800	1985	1985	2125	2125	1985
Ladefläche . . . . ca. m²	3,96	3,16	4,0	3,0	5,22	5,0	7,0	5,3	7,4	7,0
Laderaumhöhe . . . mm	1250	1170	1250	1165	1250	1430	1430	1430	1430	1430
Ladehöhe v. Boden mm	940	665	940	670	955	1000	1000	1000	1000	1000
Gewichtsdaten			Pritsche	Kasten						
Fahrgest.-Eigengew. kg	840		900	900	1070	1483	1500	1510	1565	1525
Fahrgest.-Tragfähigk. kg	1430		1600	1600	2130	2717	2900	2790	2835	3375
Zul. Vorderachsdruck kg	760		735	765	890	950	1050	950	1025	950
Zul. Hinterachsdruck kg	1510		1765	1735	2310	3250	3350	3350	3375	3950
Zul. Gesamtgewicht kg	2270		2500	2500	3200	4200	4400	4300	4400	4900
Fassungsvermögen										
Kraftstofftank . . . ltr.	57		38		57	62	62	62	62	62
Motor . . . . ltr.	3,25		5,0		5,22	6,8	6,8	6,8	6,8	6,8
Wechselgetriebe . . ltr.	1,0		0,7		1,10	1,5	1,5	1,5	1,5	1,5
Hinterachse . . . . ltr.	2,25		2,50		2,50	4	4	4	4	4
Kühlsystem . . . . ltr.	9,0		9,5		10,83	11,6	11,6	12,8	12,8	12,8
Verteilergetriebe . ltr.	—		—		—	—	—	—	—	—

[1]) Bei Einfachbereifung
[2]) Normalübersetzung ohne Stützlager
[3]) Normalübersetzung mit Stützlager
[4]) Gebirgsübersetzung ohne Stützlager
[5]) Gebirgsübersetzung mit Stützlager
[6]) Autobahnübersetzung

3,5-34 2½ t	3,5-57 2½ t	3,5—83	3,5—36	3,5—47	3,6—36	3,6—42	3,6—47	6700 Typ A
1934—37	1934—37	1934—37	1936—37	1936—37	1937—45	1937—45	1937—45	1940—45
6 79,38 117,48 3417 64 5,2	6 79,38 117,48 3417 64 5,2	6 79,38 117,48 3417 64 5,2	6 79,38 117,48 3417 64 5,2	6 79,38 117,48 3417 64 5,2	6 90 95 3626 75 6	6 90 95 3626 75 6	6 90 95 3626 75 6	6 90 95 3626 75 6
5,56 3,70 1,80 1 — 6,50	5,56 3,70 1,80 1 — 6,50	5,56 3,70 1,80 1 — 6,50	7,84 4,82 2,71 1,58 1 7,39	7,84 4,82 2,71 1,58 1 7,39	7,84 4,82 2,71 1,58 1 7,39	7,84 4,82 2,71 1,58 1 7,39	7,84 4,82 2,71 1,58 1 7,39	7,84 4,82 2,71 1,58 1 7,39
— —	— —	— —	— —	— —	— —	— —	— —	1,08 1,93
35:6 [2] 34:5 [4] 40:7 [3] 41:6 [5] 41:8 [6]	35:6 [2] 34:5 [4] 40:7 [3] 41:6 [5] 41:8 [6]	35:6 [2] 34:5 [4] 40:7 [3] 41:6 [5] 41:8 [6]	— — 40:7 [3] 41:6 [5] 41:8 [6]	— — 40:7 [3] 41:6 [5] 41:8 [6]	— — 40:7 [3] 41:6 [5] 41:8 [6]	— — 40:7 [3] 41:6 [5] 41:8 [6]	— — 40:7 [3] 41:6 [5] 41:8 [6]	40:7 Vorder- und Hinterachse — —
6,00—20 6,50—20 5″—20 5″—20 4,25 5,00	6,00—20 6,50—20 5″—20 5″—20 4,25 5,00	6,00—20 6,50—20 5″—20 5″—20 4,25 5,00	7,00—20 7,00—20 5″—20 5″—20 3,25 3,75	190—20 190—20 6″—20 6″—20 3,25 3,75	7,00—20 190—20 7,25—20 7,25—20 190—20 7,25—20 5″-20 6″ 20 6″—20 6″—20 4,0 3,5 4,0 5,3 4,2 5,3	7,00—20 7,25—20 5″—20 6″—20 4,0 5,3	190 20 7,25—20 190—20 7,25—20 6″—20 6″—20 3,5 4,0 4,2 5,5	190-20 7,25-20 190-20 7,25-20 6″—20 6″—20 3,5 4,0 4,2 5,3
6 90	6 90	6 90	6 90	6 90	6 oder 12 90 oder 50	6 oder 12 90 oder 50	12 50	12 50
3410 1433 1620 14,4	4000 1433 1620 16,5	4650 1433 1620 (1524) [1] 20 l, 18 r	3600 1545 1620 13,1	4650 1545 1620 15,4 l, 16,4 r	3600 1555 1620 (1642) [7] 13,1	4200 1555 1620 [7] (1642) 14,5	4650 1542 1620 [7] (1642) 16,5	3450 1630 1642 15,7
5380 2291 2570	6380 2291 2580	— — —	6020 2265 2660	— — —	6020 2265 2520	7025 2265 2510	— — —	5950 2340 3180
2500 2125 7,3 1430 1000	3500 2125 7,4 1430 1000	— — — — —	3495 2125 7,43 1430 1070	— — — — —	3500 2125 7,45 1430 1090	4450 2125 9,46 1430 1080	— — — — —	3500 2125 7,45 1658 1235
1530 3370 950 3950 4900	1585 3415 1100 3900 5000	1630 3470 1420 3680 5100	1815 3885 1550 4150 5700	2040 3960 1600 4400 6000	1800 4000 1700 4800 5800	1825 3975 1700 4600 5800	1900 4100 1700 4600 6000	2100 4000 1600 bis 2000 4600 6100
62 6,8 1,5 3 12,8 —	62 6,8 1,5 3 12,8 —	90 6,8 1,5 3 12,8 —	82 6 2,5 3,2 10,5 —	86 6 2,5 3,2 10,5 —	82 4,75 2,5 3,55 14,50 —	82 4,75 2,5 3,55 14,50 —	86 4,75 2,5 3,55 14,50 —	92 4,75 2,5 1,90 14,50 1,65

[7]) Nur für Hinterachsausführung mit Achsantrieb in einem besonderen abschraubbaren Hinterachsgehäuse

Displays of Opel Military Vehicles

(without guarantee of completeness)

Auto&Technik Museum Sinsheim
Obere Au 2, D-74889 Sinsheim, Germany
(3-ton standard truck 1915,
various Opel Blitz & Kapitän – 1939)

Auto&Technik Museum Speyer
Geibstrasse 2, D-67346 Speyer, Germany
(various Opel vehicles in special exhibits,
"Biber" U-Boat)

Victory Memorial Museum
Aire Victory, Europastr. E411-E25
B-6780 Messancy, Belgium
(Opel Super-6 and other cars, Blitz and Panzerwerfer)

Bibliography

Periodicals:	Continental Echo 1911-1940
	Opel-Geist 1930-1934
	Opel-Werksgemeinschaft 1934-1935
	Opel-Kamerad 1936-1941
	Modell-Magazin 1976
	Motor-Schau 1938-1944
	Motor Klassik 1984 to date
	Alt-Opel-Magazin & Markt 1972 to date
Various Data:	Opel, Miesen, Mercedes
	Private archives, including my own
Bartels, Eckhart	Opel im Kriege
Frank, Reinhard	Lastkraftwagen der Wehrmacht
Kugler, Anita	Vermögensverwaltung bei Opel 1940-42
Oswald, Werner	Kraftfahrzeuge Feuerwehr/Sanitätsdienst
Oswald, Werner	Kraftfahrzeuge Polizei/Bundesgrenzschutz
Oswald, Werner	Kfz. der Reichswehr/Wehrmacht, Bundeswehr
Spielberger, Walter	Motorisierung d. Reichswehr 1920-1935
Scherr-Thoss, H.C. von	Die Deutsche Automobilindustrie
Schneider, H.-J.	125 Jahre Opel

Photographic material was provided by:
Adam Opel AG, Bartels archives, Continental Tires archives, Guy Franz Arend, Carsten-D. Brink, Daimler-Benz-Museum, Deutsches Museum, Peter Dahlström, Karl-Heinz Denzler, Erik Eckermann, Bodo Fischer, IVECO-Magirus, Dieter Jockisch, Wolfgang Lehmann, Eckart Frhr. v. Lerchenfeld, Siegfried Maier, Carsten Messer, Opel-Maussner, Ulli Schröder, Rolf Thryller, Olaf Trapp, Hans-Martin Weber.